Brilliantly
Bad

Brilliantly *Bad*

Inventions So Terrible,
They're Good

MARK TANNER

HarperCollins*Publishers*

HarperCollins*Publishers*
The News Building
1 London Bridge Street
London SE1 9GF

www.harpercollins.co.uk

HarperCollins*Publishers*
1st Floor, Watermarque Building, Ringsend Road
Dublin 4, Ireland

First published by HarperCollins*Publishers* 2022

1 3 5 7 9 10 8 6 4 2

© Mark Tanner 2022

A catalogue record of this book is available from the
British Library.

ISBN 978-0-00-855865-9

Printed and bound in the UK using 100% renewable electricity
at CPI Group (UK) Ltd

This book is produced from independently certified FSC™ paper
to ensure responsible forest management.

For more information visit: www.harpercollins.co.uk/green

Contents

Introduction

When I was growing up in Liverpool, my dad was constantly making things. While everyone was converting to gas central heating, my dad installed our own coal-fired boiler and heating system and built a coal shed the size of a small garage next to the house. As it was my mother who had to load and stoke the boiler each morning, their marriage never quite recovered.

Some of the things my dad made, such as a coffee-pot stand fashioned from an old coat-hanger, I still have; others, such as the bird-feeding table designed to sit on top of our rotary washing line, I sadly don't.

Brought up in such an environment, I quickly developed an appreciation of and affection for a particular type of inventor's mindset. Thus inspired, and finding myself with too much time on my hands twenty-five years ago, I spent week after week at the UK Patent Office, manually trawling through tens of thousands of patent applications. The result was *Great British Inventions,* a book, published in 1997, containing a selection of some of Britain's not so great inventions.

During another bout of enforced idleness, this time due to Covid and the recent lockdowns, I was inspired to carry out a second period of patent searching, only now I was able to trawl through the online records of the United States Patent and Trademark Office. The result was the material for *Brilliantly Bad.*

If you ever have a great idea for an invention, such as the Toilet Seat Volatile Gas Incinerator that features on page 170 of this book, then the patent system is there to help ensure that no else can make or sell your invention without your agreement. The process can be long and expensive, but if you believe in the potential of your idea, then filing a patent application is an essential step. Each year, in a testament to human ingenuity, over 3 million patent applications are filed worldwide.

In applying for a patent, the inventor needs to describe the thinking that led to the invention and the particular problem it solves. What problem, for example, does a Combined Aquarium and Cat Display Case seek to address and how? The applicant also needs to provide a description of their idea and how it works, along with accompanying drawings.

In selecting the inventions for this book, I have focused on those where inventors have sought to enhance fundamental areas of our lives. For men, for example, this addresses such critical issues as hair (or lack of), failings in the area of personal hygiene, and, of course, the size, shape and vitality of the penis.

In some instances, different inventors tackle the same problem but come at it from very different angles. Solutions to the universal problem of splashes caused by the male inability to urinate accurately will depend, for example, on whether the man is attempting to urinate with a limp or semi-erect penis. Each situation has inspired distinct and innovative solutions.

All the inventions in this book have been submitted and published as official patent applications. Some may already be in production; others, like my dad's bird table, may not.

For Men

Magnetic Hairpiece

Keep your hairpiece from blowing off by having magnets implanted in your scalp.

It is a fact of life that many men around the world suffer from partial or even total baldness. As many people view such baldness as undesirable, a flourishing industry has arisen devoted to solving this problem.

One of the most common ways of disguising baldness is through the use of an artificial hairpiece. Usually, the bald or balding man wears this on his head to create the illusion of a full head of hair. A problem can arise, however, in high winds or during periods of vigorous physical activity. In such situations, the bald or balding man may have to deal with the highly embarrassing possibility that his hairpiece will come loose, or even fall off.

The usual method of preventing such embarrassment is to glue the hairpiece to the scalp. This can, however, be messy and also cause allergic reactions. A better means of securing a hairpiece to the head is therefore desired.

The present invention provides such an improvement. The method involves implanting magnets beneath the skin of a bald head and then magnetically securing a hairpiece to the implanted magnets. This holds the hairpiece to the head firmly, but temporarily.

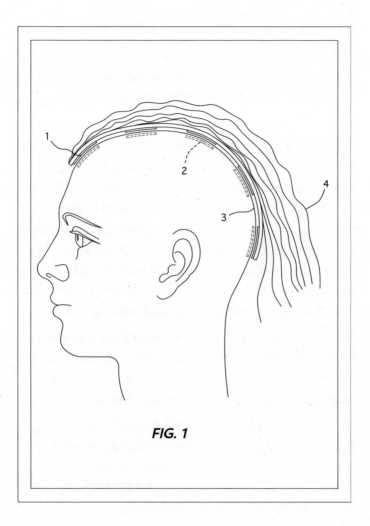

FIG. 1

Combined Coat and Urinal

Urinate in public without anyone else knowing.

The dearth of public toilets, especially in urban areas, has been the cause of significant distress to untold numbers of people who have been in need of such facilities.

As a result, there has been a considerable rise in public urination, along with a consequent degrading of city life. It is therefore the aim of the current invention to provide a garment that enables a man to relieve himself inconspicuously and sanitarily in a public situation.

The invention is comprised of a conventional coat which has specially adapted sleeves with simulated hands sewn to the cuffs. The simulated hands make it appear that the wearer of the coat has his own hands at his side outside the garment. In fact, because the sleeves of the coat have concealed slits, the wearer can use his real hands to access a hidden pocket inside the garment which holds a container for collecting urine.

Should no public toilet be available when relief is required, a man can make his way to an inconspicuous place, don the garment and safely relieve himself without the risk of becoming a public nuisance. To anyone who is not too near, he would simply appear to be a pedestrian casually standing and observing his surroundings.

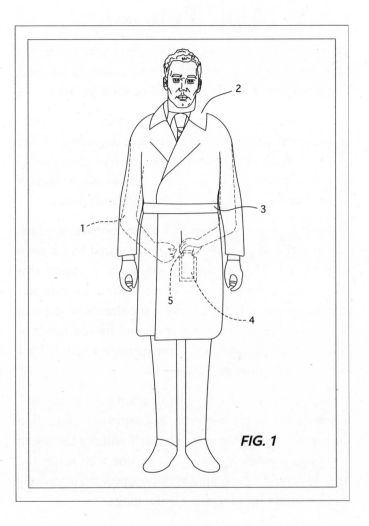

FIG. 1

Device for Measuring Male Potency

Gauge the size of your erection at any time of day or night.

It is generally known that a typical male has three or four spontaneous erections of varying magnitude during a normal night's sleep. Clearly, as he is asleep, he cannot monitor this. It is therefore the aim of the present invention to provide a cheap, simple and self-administered device to make this possible.

Previous methods to measure the tumescence of nocturnal erections have proved cumbersome. Additionally, experience has shown that at least three nights of testing are needed to produce reliable results. This can be time-consuming, sleep-depriving and expensive. The present invention provides a simple, cheap, easy-to-use alternative.

In use, the closed loop is passed over the end of the limp penis and tightened to fit. To hold the loop in place, a piece of 'Band-Aid' can be used. If an erection occurs, the tail slides partially out of the loop to accommodate the amount of penile expansion. It remains in this position until it is manually removed in the morning. The size of the erection can then be assessed by measuring the extent of the tail that is left exposed.

The invention can clearly also be used when the person is awake.

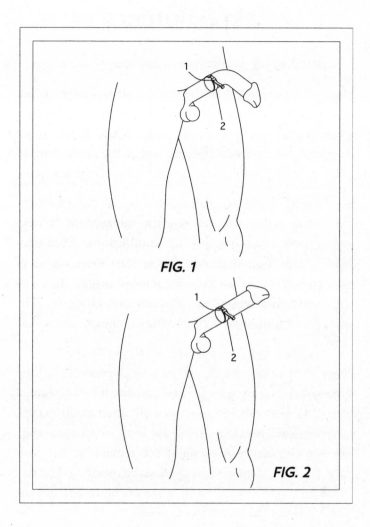

FIG. 1

FIG. 2

Automatic Hair-cutting Apparatus

Ensure you get exactly the same haircut every time.

The present invention relates to a machine for mechanically cutting hair on the head of a man (or woman) whereby a complete haircut is obtained in a single operation without requiring any manual cutting of the hair.

The person whose hair is to be cut first positions their head directly under the hair-cutting helmet. The helmet is then lowered so that the head can be measured as to size and shape. A plurality of individual mechanically driven cutters mounted in the helmet are then positioned in accordance with the style of haircut desired. The cutters operate simultaneously to rapidly cut the hair to the proper length over the entire surface of the head.

The helmet is additionally connected to a vacuum blower to carry away the cut hair, while ear plungers can be used to keep hair out of the ears and also to provide a noise barrier. In one embodiment, the helmet is provided with conductors for creating a static electric field to cause the hair to rise upwardly to facilitate cutting.

The person's head measurements and desired hairstyle can be stored in the system for future use, guaranteeing the same haircut every time.

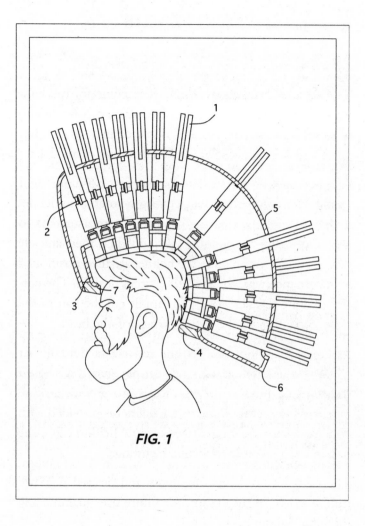

FIG. 1

Sweat Guard

Protect your shirt collar and hat with a panty liner.

A crisp, spotless shirt collar is absolutely essential if one wishes to convey an image of confidence, professionalism and success to one's fellow man. A smart, well-kept hat is another prerequisite. Sweat stains seriously detract from such an image.

Sweating around the neck, which is a perfectly normal human function, can cause unsightly stains on the individual's shirt collar. Likewise, haircare products can mix with sweat to create embarrassing stains on the inside of one's hat.

The aim of the current invention is to prevent such undesirable effects. The basic principle is to readapt an existing panty-liner system to a create a sweat collection and stain prevention system for shirt collars and hats. Rather than being used to line the crotch area of an undergarment, the moisture-absorbent panty-liner material is attached to the interior of a shirt collar or hat.

As the adapted panty liner is worn on the inside of a collar or hat, it is not easily discernible and, being lightweight, does not create any unsightly bumps and bulges.

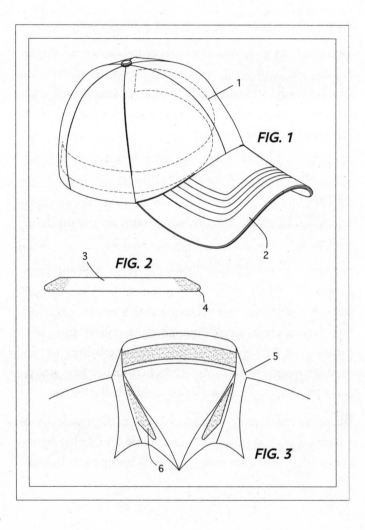

FIG. 1

FIG. 2

FIG. 3

Condom-carrying Necktie

Keep your condom conveniently, but discreetly, at the ready.

The need to be prepared for sexual activity dictates that a condom be carried on one's person at all times. The traditional method of carrying a condom in one's wallet has the problem that the condom can create a tell-tale circle on the surface of the wallet. This can lead to the user being the subject of jokes and amusement, therefore discouraging the commendable practice of being ready and prepared at all times.

Furthermore, keeping a condom in one's wallet risks subjecting the condom to maltreatment and can, at the very least, spoil the appearance of the condom packet, giving the highly unwelcome impression that the condom may be unsafe to use.

The current invention overcomes such problems by providing an aesthetically pleasing necktie with a hidden pocket designed to carry a condom, or other similar-sized item, securely and discreetly.

Preferably, the pocket remains concealed while the tie is being donned and worn.

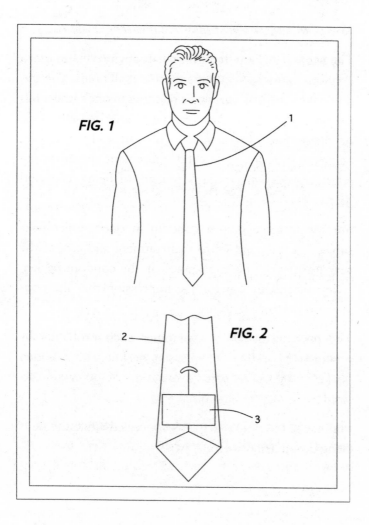

FIG. 1

1

FIG. 2

2

3

Improved Hairpiece

Make sure your hairpiece always stays firmly in place.

The present invention provides an improved hairpiece. More particularly, the improvements relate to the ease of attaching a hairpiece to the bald area of a user's head with sufficient firmness to prevent inadvertent detachment.

The invention nevertheless permits, without any complicated and tedious procedure, the easy removal of the hairpiece prior to sleep, swimming or other occasions when removal enhances the comfort of the wearer without causing any accompanying embarrassment.

The invention works by attaching an outer component to the remaining natural hair of the wearer by adhesive or other means. The hairpiece is then secured to the component by means of hinge pins of the type typically used for a pivot hinge. The pins can be inserted and the hairpiece attached when the wearer is preparing for a public situation and can be readily withdrawn when conditions of privacy prevail.

The outer component, to which the hairpiece is connected, is 'permanently' attached to the user's head and is not removable for sleep, exercising or other such occasions.

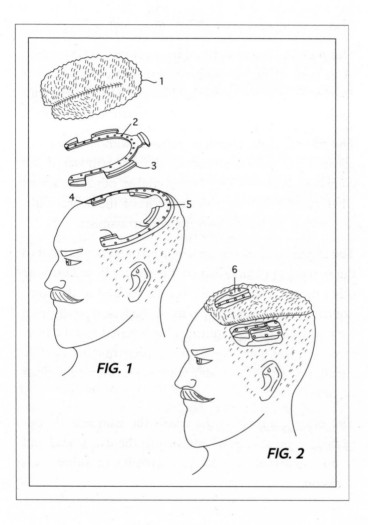

FIG. 1

FIG. 2

Device for Testing One's Breath

Smell your own breath before others do.

Many men suffering from malodorous breath are blissfully unaware of the fact. Because of this, they risk causing offence to people they meet without knowing it.

To remedy this problem, the present invention provides a means of testing one's breath which can be used quickly, simply and discreetly. A face mask made of impervious material is designed to fit securely over the mouth and nose. With the mask in place, the user exhales through the mouth into the mask, which is designed to retain a substantial quantity of the exhaled gases. Keeping the mask in place, the user then breathes in the exhaled gases through the nostrils, allowing him to smell the possible malodorous nature of his breath.

Should the initial test be indecisive, one or two additional breathing cycles may be undertaken without removing the mask from the face. This will reinforce the presence of any malodorous gases emitted during exhaling, making it easier to ascertain whether the breath is fresh or not.

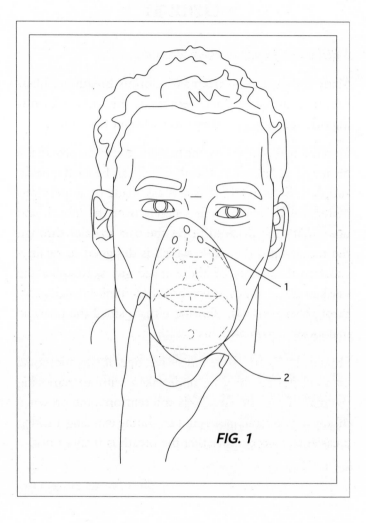

FIG. 1

Soft Goods for Eliminating Crotch Itch

End the misery caused by your testicles rubbing against your thighs.

The current invention provides a male soft goods appliance that effectively brings an end to the soreness and irritation produced by the condition most commonly known as 'jock itch'. This condition affects men of all ages and causes misery and discomfort to all those who encounter it.

The appliance consists of an elasticated waistband with a soft padded fabric strap at the back which passes downwards between the buttocks and past the anus to form a cup-like pouch to hold the scrotum. Side straps ensure that the penis is separate from the scrotum while also enabling the penis to remain free and uncovered. When worn, the central strap effectively isolates the buttocks from each other, while the pouch separates the scrotum from the thighs and stops them rubbing together.

Experience has shown that as long as the appliance is worn regularly, it can provide substantial relief from the irritation and misery caused by jock itch. Furthermore, the appliance is comfortable to wear and is also unnoticeable.

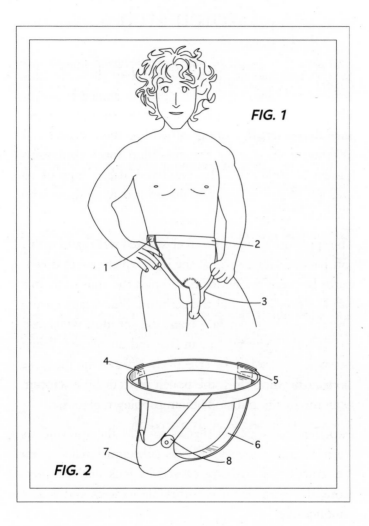

FIG. 1

FIG. 2

Sideburn-trimming Guide

Make certain your sideburns are always of equal length, even when shaving in the dark.

Smartly trimmed sideburns speak well of the wearer and contribute to a well-groomed, neat appearance. However, as many can testify, the problem of trimming sideburns to the same length and the same angle is considerable.

In the past, sideburn-trimming guides have been provided, but have offered limited adjustability and haphazard positioning.

The present invention overcomes these shortcomings. It includes a flexible band to be placed over the user's head. The band can be adjusted to fit heads of various shapes and sizes. Positioning elements rest on the user's ears, and adjustable guide plates, providing a straight guide edge for the user's razor, are secured to the ends of the band.

A built-in lamp is used to illuminate the sideburns adjacent to the guide. The lamp is activated when the sideburn-trimming guide is placed on the user's head.

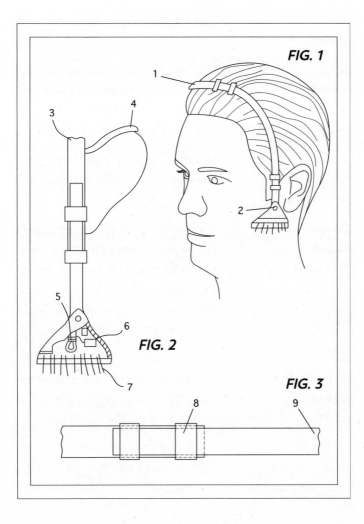

FIG. 1

FIG. 2

FIG. 3

Penis-straightening Underpants

Keep your penis straight when wearing tight-fitting trousers.

According to medical research, a considerable number of adult males experience an unwanted sideways bend in their penis. Investigations into the causes of this problem have found that it principally results from the regular wearing of tight-fitting jeans or trousers for long periods of time. If an erection occurs, this can cause compression and bending of the reproductive organ.

This is a particular problem for young men. In particular, when a male reaches puberty, because of his exuberant and impulsive nature, as well as other factors, frequent erections of the penis naturally occur. If he is wearing tight-fitting jeans or pants when an unavoidable erection occurs, the penis can be subject to considerable flexural deformation. The larger the erection, the more apparent this will be.

The present invention overcomes this problem by providing a specially designed pair of underpants which fix the position of the penis centrally to prevent it from moving to one side. Therefore, if an erection occurs, the underpants will hold the penis in place, pointing directly upwards, thus preventing it from bending.

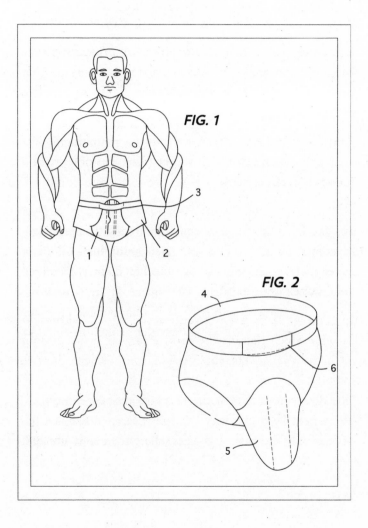

FIG. 1

FIG. 2

Clip for Protruding Ears

Stop your ears from sticking out by clipping them to your hair.

Many men suffer from the problem of protruding ears. In some instances, only one ear protrudes, and in other cases both. Either can seriously detract from a smart, professional appearance.

At present, the only effective means of flattening the ear against the head is through surgery. This has obvious and unappealing drawbacks.

As far as the present inventor knows, no one has yet designed a clip for flattening protruding ears which is comfortable, reliable and also unnoticeable. Prior attempts at such devices have produced ear clips which are cumbersome, annoying to use and highly visible.

In contrast, the present invention provides a lightweight ear clip which flattens the protruding ear by securing it to the user's hair. The clip is made from resilient, durable material, such as a spring steel core, covered with plastic. The design is such that the clip is practically unnoticeable, especially if the user carefully drapes a few strands of hair over the part of the ear where the clip is situated.

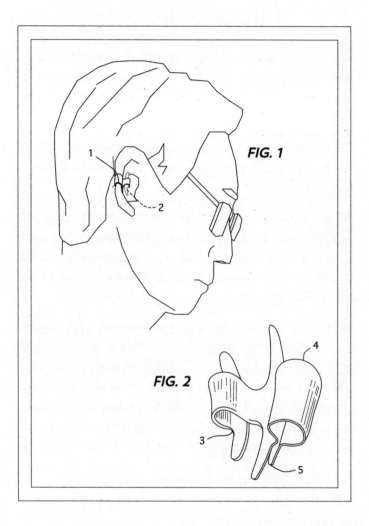

FIG. 1

FIG. 2

Method of Concealing Partial Baldness

Save a small fortune hiding your bald patch with this advanced comb-over method.

For men who are partially bald and wish to disguise the bald area, hair transplants, hair weaving and hairpieces are the most common solutions. The cost of these methods, however, can be highly expensive.

Obviously, a partially bald man without the necessary financial means cannot afford such luxuries. He is, therefore, left with few options. One method is to use his own hair to cover the bald area, but, as a rule, most people lack the ability to properly execute a hairstyle of this type that will look good.

The method herein disclosed provides a solution. It involves using the remaining hair around a bald person's head as a covering. This hair must first be allowed to grow extra-long, and is then brushed over the bald area in alternating folds, using hairspray to hold each section in place. The final top section can then be styled to the person's personal taste.

By lightly sweeping the hair into the desired style as the hairspray dries, an appearance of a full head of hair is given, as in Fig. 6.

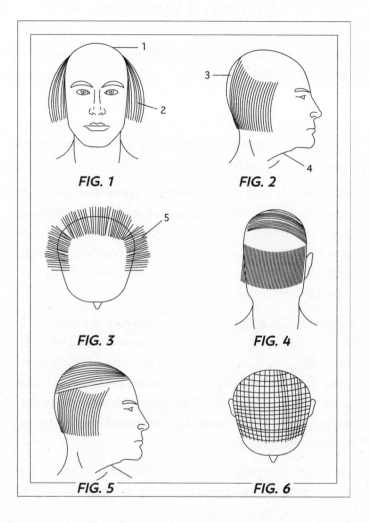

FIG. 1

FIG. 2

FIG. 3

FIG. 4

FIG. 5

FIG. 6

Sleeping Device for a Sitting Position

Fall asleep in comfort while sitting down.

Many people, particularly men, have a tendency to fall asleep in all manner of places, not least when they are sitting down. To make such sleeping more comfortable, the present invention provides a device upon which a person can rest while in a sitting position.

The device includes a rounded depression to accommodate a sponge on which the person can comfortably rest their chin. It also provides a U-shaped channel in which to comfortably rest both arms, and two rounded grooves in the lower portion that, when placed on the lap, fit on both legs.

As well as a resting device, the invention can also serve as a food tray and as a storage facility for food. It is ideal for use while seated in a vehicle, such as by truck drivers resting on long trips, or it can be used by numerous other persons in a wide variety of situations, such as at picnics, at the beach or watching TV, or by construction workers on their lunch break or the like.

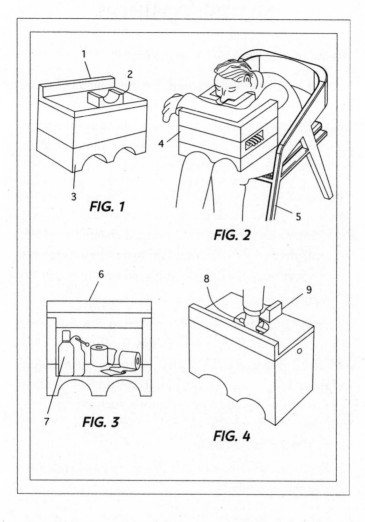

FIG. 1

FIG. 2

FIG. 3

FIG. 4

Weight-lifting Device for the Penis

Pump iron with your penis to make it bigger and stronger.

It is recommended that the following exercise programme be carried out at home:

1. Ensure you have a strong and sturdy erection.

2. Stand or sit comfortably with your feet apart. Tighten your stomach muscles and then thrust your pelvis and erect penis forward.

3. Carefully position both straps of the device on the erect penis: one strap at the base and the other near the glans.

4. Add weights to the device. Around 50g is recommended at the start of the session. Further weights can be added as the session progresses.

5. With the weights in position, start flexing the erect penis for a series of 10 lifts.

6. Stimulate the glans of the penis as and when required in order to maintain an erection.

7. Rest after 10 lifts and then increase the weights by 10g and continue.

8. Repeat the process until you are unable to hold your erection any longer.

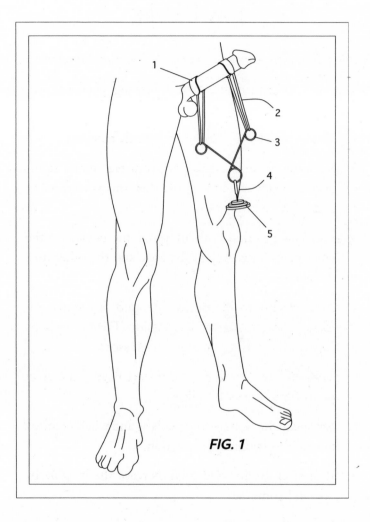

FIG. 1

For Women

Breast-development Apparatus

Unhappy with your breasts but worried about silicone implants?

The present invention relates to a novel breast development apparatus to assist all women in the enhancement of their breasts. It is intended for use by women in the privacy of their own home without requiring any special skill or expertise.

In use, the cups are placed over a woman's breasts and the handles then manipulated in a back and forth direction. This creates suction, which draws the breast into the cup, stimulating the breast and, with regular use, enhancing its size and firmness.

The apparatus has several desirable features. It is relatively simple in its construction and therefore may be readily manufactured at a relatively low price to encourage its widespread use. As it is devoid of working parts, it is unlikely to get out of order. It is also aesthetically pleasing and refined in appearance.

Fig. 1 is a view of the packaged breast development apparatus kit.

Fig. 2 is a perspective illustrating the manner of applying the present invention to both breasts at the same time.

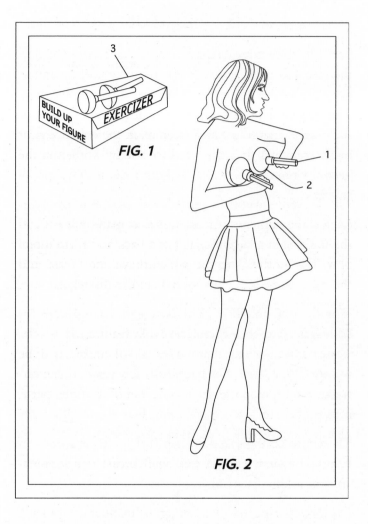

FIG. 1

FIG. 2

Face-lifting Equipment

Achieve a younger look by lifting your skin with Velcro pads.

Face lifts to remove unappealing lines and wrinkles from the face and neck generally involve a surgical operation of some kind to raise the skin around the temple area. Due to the time-consuming and expensive nature of such surgery, attempts have been made to provide equipment which will give the effect of a face lift without the need for surgery.

The present invention provides a great improvement on such equipment by utilising Velcro pads that stick to the user's skin adjacent to her temples. As the pads pull upwardly on the skin, the jowls are instantly lifted and the unwanted neck wrinkles and the like disappear.

Moreover, a unique Velcro fastening arrangement, in combination with a hairpiece overlying the top of the woman's head, means that the tension of the pads can be very easily adjusted. The hairpiece itself not only camouflages the fastening arrangement, but adds more thickness to the hair.

The entire arrangement is such that the equipment can be used by the woman herself without requiring the services or assistance of anyone else.

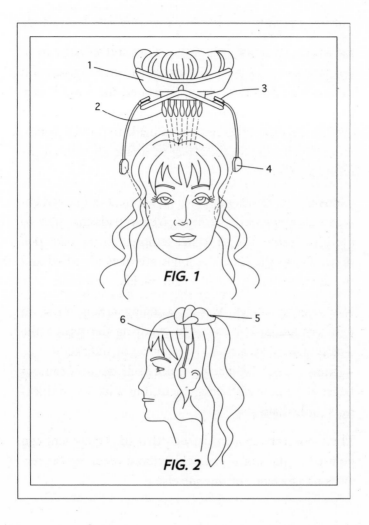

FIG. 1

FIG. 2

Slimming Mirror

See how a slimmer you might look.

Mirrors have traditionally been provided to give a relatively true reflected image of an object. Alternatively, mirrors have also been provided which distort the object.

In one invention, a mirror was designed to show a person how they would look after losing a certain amount of weight. This invention was deficient in that it provided an image of the person's weight loss over both the body and the head. This could give a very misleading appearance, as a reduction in the dimensions of the face following weight loss is usually much less than the reduction in the size and shape of the body.

It is therefore an object of the present invention to provide a mirror in which the appearance of the viewer's face will be unchanged, while the image of their body will be altered. To achieve this, the top part of the mirror remains planar and does not distort the head, while the lower portion is curved, so as to give an impression of the body after weight loss (or gain).

Fig. 1 is a perspective view of the mirror being used by an overweight person.

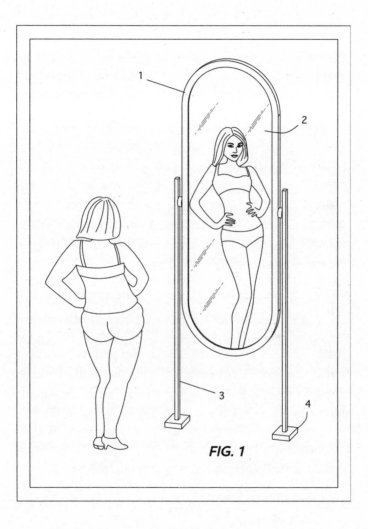

FIG. 1

Combination Beachwear and Seat Cover

Wear your car seat cover as an attractive piece of bathing attire.

People planning to visit the beach often prefer to don their bathing suit in the privacy of their home and then drive to the beach. This causes a problem, however, as there is a need to protect the vehicle seat from perspiration, as well as providing a way to more fully cover the body before reaching the beach.

To avoid the undesirable consequences to the car seat brought about by travelling in beachwear, various make-shift expedients have previously been tried. These typically involve draping some kind of towelling over the vehicle seat. This, however, gives a false sense of security, as the towelling cannot be securely attached to the seat.

In view of the foregoing, the present invention provides a unique and versatile item which serves both as a protective car-seat cover and as a stylish article of beachwear for lightly clad bathers. The item can first be used as a seat cover by the person travelling to the beach and then as an attractive piece of bathing attire on the beach.

Later, should it be wished, it can also be used as a cover for a beach chair.

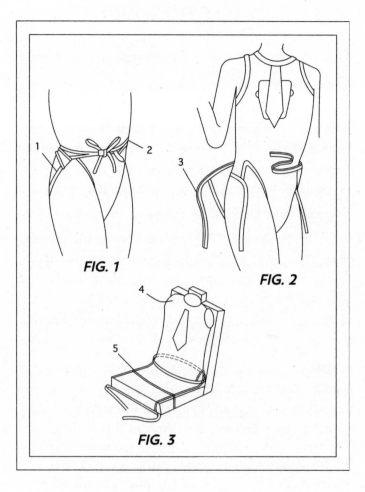

FIG. 1

FIG. 2

FIG. 3

Apparatus to Reduce Breast Wrinkles

Stop your breasts wrinkling when you sleep.

As women get older, lose weight or become less active, their breasts have a tendency to become less firm. Additionally, if a woman sleeps on her side, her breasts can rub together, causing wrinkles to form in the cleavage area.

Accordingly, the current invention provides an apparatus designed to reduce the development of such wrinkles when the wearer sleeps on her side.

The apparatus consists of a cylindrical, semi-firm material, such as rubber, with an optional outer casing of washable satin. The satin casing may additionally have lace or other decorative material to enhance its aesthetic appeal.

In use, the invention sits comfortably between the wearer's two breasts and can be secured in place either by the wearer's sleeping apparel or by a special belt if the wearer prefers to sleep nude.

By keeping the two breasts separated, the pull and stretch caused by one breast hanging over the other as the wearer sleeps is reduced, thereby preventing future wrinkles from forming and also softening and smoothing existing wrinkles.

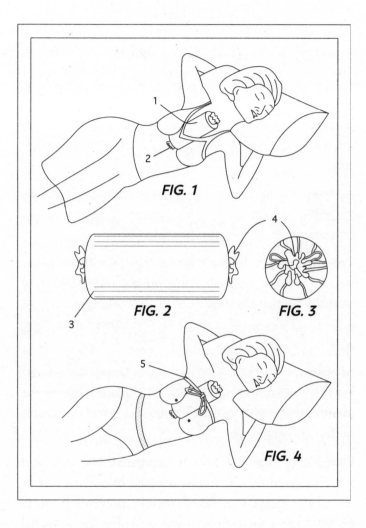

FIG. 1

FIG. 2

FIG. 3

FIG. 4

Arms-free Flotation Device

Enjoy the water in safety and style.

Various flotation devices have long been used for recreational purposes in swimming pools, oceans, lakes or other bodies of water. Such devices have generally been tube-shaped products or life jacket and vest-type devices. The disadvantage of such devices is that they can limit the user's arm movement when engaging in water-based recreational activities.

The present invention overcomes this problem. The crotch portion of the garment is positioned between the user's legs, with the front portion in contact with their abdomen and the back portion in contact with their buttocks and lower back. It is held in place by straps with buckles, snaps or Velcro.

Due to the configuration of the garment, the user is floated in a comfortable, relaxed manner, either upright or seated, with their head safely above the top of the water and their arms completely free for paddling or for playing water polo or other games.

The garment can be made in various sizes to fit not only the dimensions but also the weight of the user.

FIG. 1

Beauty Paddle

Keep your skin looking soft, young and supple.

The idea of slapping the skin to improve circulation is prehistoric and probably began as a response to shock, freezing, drowning or the like. In 'primitive' cultures, slapping the skin to alter appearance is widespread.

Until the present invention, however, devices designed to facilitate such slapping have all been unpowered, manual swatters. The present invention overcomes the attendant problems of such devices by providing a multiple-bladed paddle with an electric motor. A soft paddle of a rubber or plastic composition would probably be most sympathetic to the skin.

In use, the paddle may be held near to the face or further away and the speed adjusted to the desired level of stimulation. Further explanation is believed superfluous.

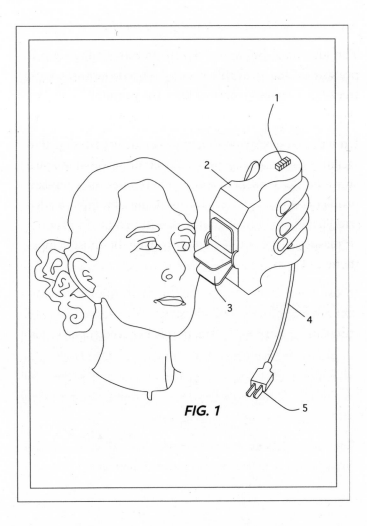

FIG. 1

Buoyant Bathing Brassiere

Relax in the water, knowing that you won't sink.

This invention relates to brassieres which may be worn by water-skiers and bathers to render them buoyant and to protect them against submersion in water.

The brassiere is made buoyant by the addition of strips of a buoyant filler material to the cups. Sufficient buoyant material can be incorporated in the garment to render buoyant even those persons known as sinkers – that is, persons not having the usual amount of body buoyancy. The garment is positioned so that even if a wearer should be rendered unconscious following a fall in water-skiing or the like, their body will rotate to assume a position in the water with the brassiere and face uppermost to protect against drowning.

The brassiere is easily donned and removed and is held in place effectively by shoulder straps and the back band. It is designed to fit the wearer snugly and permit full freedom of movement when swimming, water-skiing and engaging in other activities.

The cup portions have a suction-like adherence to the breasts which resists displacement when worn.

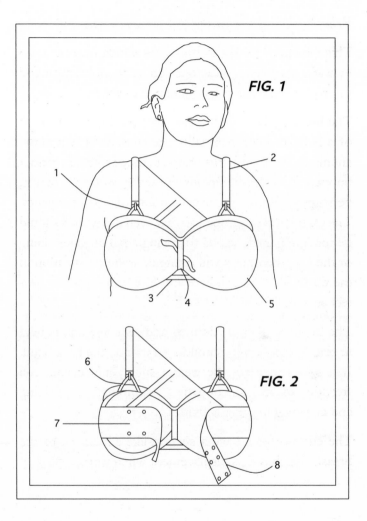

FIG. 1

FIG. 2

Face-lifting Earrings

Keep your earlobes pinned back to smooth out your facial wrinkles.

The onset of ageing and loss of youth are evidenced most clearly in the slow sagging of facial tissue. Face creams, oils and ointments promising to improve this have become a multi-billion dollar a year industry. Additionally, a range of invasive methods, including surgery, are used to stretch the skin and return it to a more youthful appearance.

As yet, however, no device currently exists to smooth out the wrinkles of the face and chin using earrings. The current invention is such a device.

The process used is reasonably non-invasive and only requires two piercings to be made in each ear: one on the earlobe and another a little higher up. The earlobe is then bent back behind the ear so that the first hole aligns with the second hole. By folding the earlobe in two and then securing it back with the earring, any wrinkles and loose skin on the face are smoothed out by the pull of the earrings.

A decorative ornamental cover can be used to obscure the folded earlobe along with any scrunched-up skin.

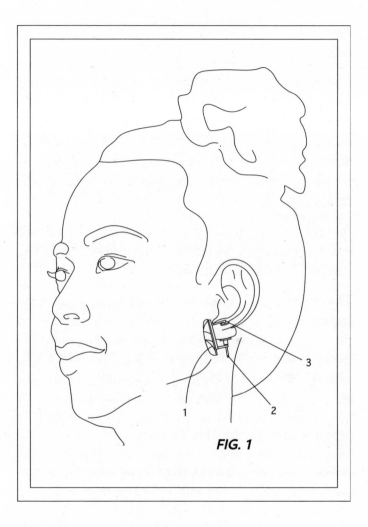

FIG. 1

Feminine Undergarment with Calendar

Calculate when your next period is due using the calendar on your knickers.

Conventionally, various methods have been used in order to work out the likely starting date of a woman's next period. Such methods include remembering the date of the last period and computing the cycle to determine the starting date, or recording previous dates in a notebook and then working out the next date.

However, it is not only difficult but also quite complicated to remember the date of the last period and then compute the cycle. It can also be extremely bothersome to regularly record the results. This invention solves such troublesome problems.

It provides a feminine undergarment with two heart-shaped clips and a calendar. The first clip is used as a pointer for the dates from the 1st to the 15th day of the month. The second clip is used as a pointer for the dates from the 16th to the 31st. Either of the clips may be set to show the date of the onset of the period of the wearer, from which she can easily predict the date of her next period.

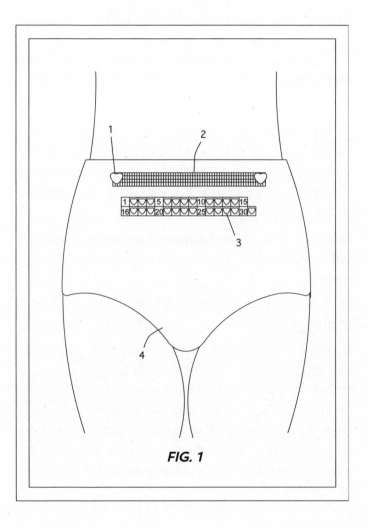

FIG. 1

Light-emitting Underwear

Ensure that your bra and knickers will be appreciated, even in the dark.

In the current fashion, ladies' underwear, such as knickers and brassieres, for example, are well designed to keep the attention. Such designs are concentrated on colours, shapes or decorations to create novel and appealing effects. In the dark, however, it is not possible to see the beautiful outlooks of these designs, even though it is well known that ladies' underwear is usually displayed in the dark. Thus, there is a long-standing demand for a design to make better the defects of previous knickers and brassieres.

The underwear invention does this by providing a pair of light-emitting knickers and a light-emitting brassiere so that this underwear can present beautiful light effects.

Referring to a pair of knickers as an example, they are fitted with an edge strip upon which there is a plurality of light-emitting elements connected to a circuit board installed with batteries and control switches. It is only necessary to press the switches to turn on the lighting elements, thereby presenting a beautiful effect on the knickers.

The same effect can be created in a similar way on the light-emitting brassiere.

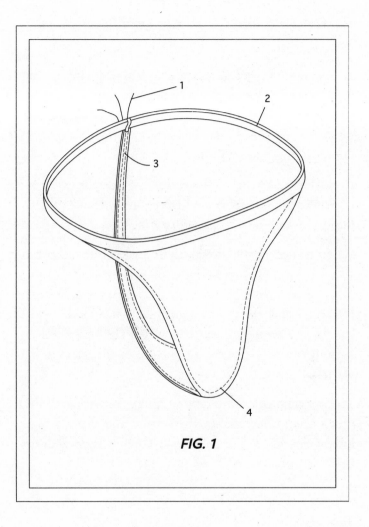

FIG. 1

Male Y-front Top

Surprise your boyfriend by sporting his Y-fronts as a funky top.

In this new age of funky clothing, it appears there is no limit to the imagination of fashion lovers. Women especially seem to have a fascination for wearing clothing that was originally designed to be worn by men, or at least designed as if it were to be worn by men.

Accordingly, the present invention provides a novel item of female apparel fabricated from, or fabricated to have the appearance of, a man's tight-fitting briefs, or Y-fronts.

When turned upside down, these male underpants can be worn over the upper torso of a woman as a funky top. The leg openings serve to receive the female wearer's arms, while the crotch is adapted to include an opening for the wearer's head and neck. The characteristic Y-front can be situated on the chest or the back of the wearer as desired.

The appearance of the tightly fitting, somewhat elastic, men's briefs will enhance the slender and svelte female look which has in recent times become extremely desirable and popular.

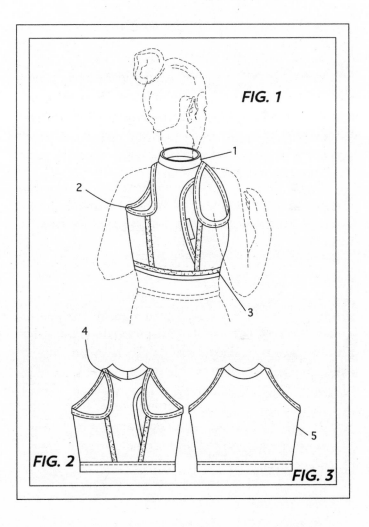

FIG. 1

FIG. 2

FIG. 3

Nipple Suppression Device

No longer feel the need to hide your nipples with crossed arms or a notebook.

For many women, an immaculate appearance is crucial to their feelings of confidence and self-assurance. Regrettably, for a number of women, this confidence can be undermined by the outline of their nipples creating unwanted bumps which interrupt the smooth lines of their form-fitting clothing.

To conceal these bumps, some women walk around with their arms permanently crossed in front of their chest. In an office situation, other women will never leave their desk without holding a notebook, a stack of papers or a file folder close to their chest. The present invention is designed with these women in mind.

The device is comprised of a soft, smooth nipple cover held in place by a silicon ring. It can be worn with all types of figure-hugging clothing, from evening wear to swimsuits. The ring can be made in a variety of sizes to fit women with either very large nipples or very small nipples. However, testing has found that a single-size device is able to accommodate the nipples of a surprisingly large number of women.

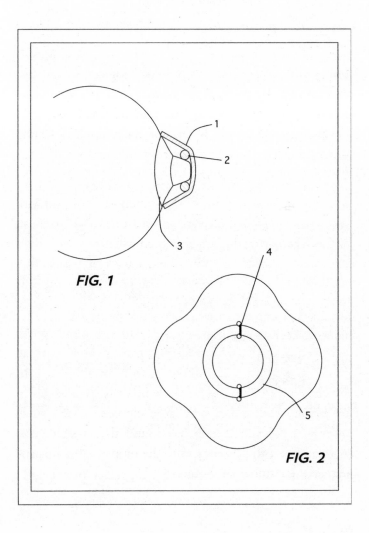

FIG. 1

FIG. 2

Pubic Patches

Change your pubic hairstyle quickly and easily.

The removal by Western women of parts of their pubic hair has become increasingly common since around 1945, when bathing suits became more diminutive. Since then, a variety of pubic styles have been adopted, including the American, French, Playboy, Brazilian (half and full) and Sphinx.

However, a woman who wishes to regularly change her pubic hairstyle can find this a painful and expensive experience. There is therefore a need for a product that enables women do this cheaply and easily.

The present invention meets this need. It provides a variety of different 'pubic patches' with an adhesive on one side and 'hair' on the other. The side with the adhesive is then stuck on the pubic area.

The patch can be of various shapes and sizes, and the hair can be of different colours – redhead, blonde, brunette, etc. The hair can also range in stiffness from hard to soft, and can be curly, straight or bushy.

The patches can be made of a variety of materials, including natural or synthetic hair, fur, velvet, plastic and feathers.

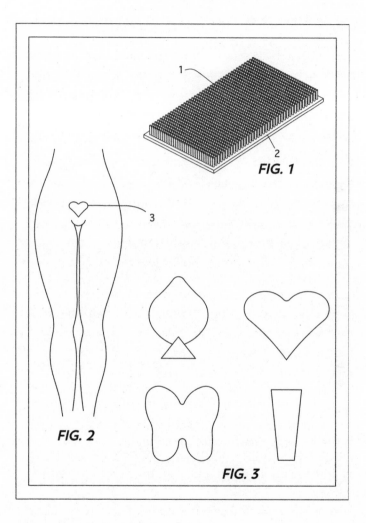

FIG. 1

FIG. 2

FIG. 3

Shoulder Strap Retainer

Prevent your shoulder strap from slipping by sticking it to your shoulder.

Shoulder bags have become an increasingly popular fashion accessory. However, the tendency of the strap to slip off the shoulder can make them particularly annoying to wear.

Previous solutions to this problem have involved providing retaining strips which hold the shoulder strap in place by attaching it to the garment that the person is wearing. This, however, either requires alterations to the garment or creates unattractive holes therein. Furthermore, such solutions are clearly unfeasible when the person is wearing a strapless swimsuit or other strapless garment.

With this in mind, the present invention provides an alternative method of securing the shoulder strap to the shoulder. This is achieved by gluing a special strap retainer to the skin of the wearer by means of an adhesive typically used in applying medical dressings to the skin.

Both the strap retainer and the shoulder bag strap are fitted with two-part snap fasteners so that once the retainer is glued to the shoulder, the strap can be snapped in place.

It should be noted that other types of fasteners, including Velcro strips, for example, may also be used.

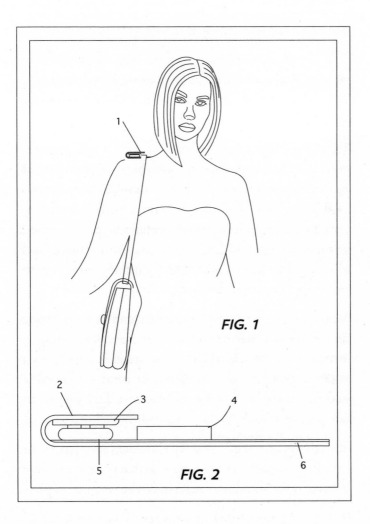

FIG. 1

FIG. 2

Zip Operating Mechanism

Never have to say, 'Can you do up my zip?' again.

There are many challenges women face during the day, such as getting dressed. In particular, they can have a challenging time opening and closing zips on garments, especially garments with a zip at the back or side.

Operating a zip on the back of a dress can be difficult, whether it be in a woman's own home or at any other time when she is on her own and wants to be independent. Indeed, research indicates some women spend an average of nine hours a year struggling to open and close their garments. In addition, every woman wants to feel confident that her zip will not accidentally open due to not being completely done up.

The present invention therefore provides a zip operating mechanism to assist women in this process. The mechanism is easily mounted over door ends and includes a system of pulleys and an endless cord to which a hook is attached. In use, the hook is secured to the tab on the zip fastener and a pull on the cord will move both pulleys, resulting in the zip being raised.

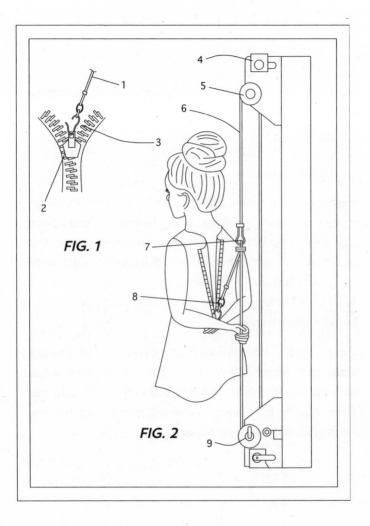

FIG. 1

FIG. 2

Swimsuit with Novel Attachment Means

Keep your swimsuit plugged securely in place.

People wear clothing for three main reasons: for decency, to protect themselves from their surroundings, and to keep abreast of the changing winds of fashion. To accomplish any of these objectives, it is essential that the garments do not fall off the body.

Until now, there have only been four known methods to ensure the body remains clothed: wrapping, clamping, draping and gluing. The present invention presents an innovative fifth method – one that is especially suited to a swimsuit for a woman who may desire to expose her entire backside, but still, at least minimally, cover her breasts and genitals.

An additional benefit of this invention is that it allows the wearer to remove the entire garment without having to lift either leg, thus minimising the risk of falling over. It also precludes having to lean on something, or someone, while removing or replacing one's swimsuit.

The design also permits the wearer to store a few small objects (e.g., a key, ID, small change) inside the plug to keep them dry, even while swimming.

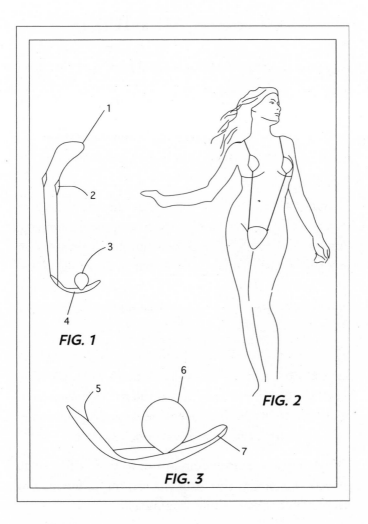

FIG. 1

FIG. 2

FIG. 3

For Pets

Animal Ear Protectors

Stop your pet's ears from dangling in their food bowl when they eat.

The present invention concerns itself with a problem which has long been recognised but never adequately solved: how to protect the ears of long-eared animals, especially dogs, from coming into contact with their food or drink while they eat or drink. It is a further objective of the invention to ensure the device itself does not come into contact with the animal's food and is also lightweight, comfortable and not easily removed by the animal.

The device consists of two protective ear tubes which are secured to the animal's head by two straps. The straps are easily adjustable to accommodate animal heads and ears of various sizes. This also allows for the comfortable consumption of food and drink by the animal while it is wearing the device. Furthermore, the invention itself may be decorated so as to enhance the appearance of the animal in the eyes of its owner and of others.

Fig. 1 illustrates a long-eared Cavalier King Charles spaniel wearing the described invention, but the invention should not be viewed as limited to spaniels, since it is equally useful for any animal with long ears which may dangle into its food or drink.

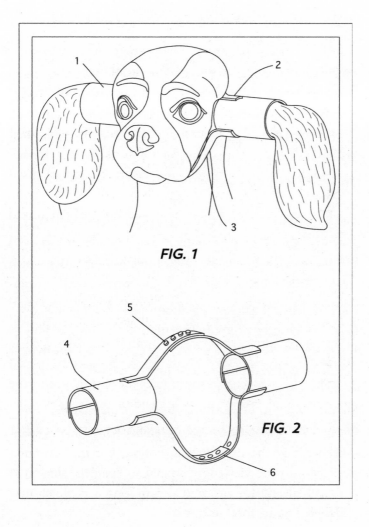

FIG. 1

FIG. 2

Animal Time Device

Appreciate how your pet experiences the passage of time.

Watches, clocks and other time-keeping devices traditionally record the passage of time in human terms. A day, an hour and a minute have a value which is measured in relation to the duration of a human lifetime: a week has greater value than a day; a year greater value still.

However, when measured in terms of human time, animals such as dogs live shorter lives than people. Traditionally, this relationship is related in 'dog years', with one 'dog year' equating to seven human years. Although this may be helpful in deciding if a dog is fully grown or not, it does little to help the owner put a correct value on their animal's time.

Accordingly, the present invention provides a watch or clock which runs at a different time from human time. For dogs, the hour hand revolves seven times faster than a human watch. For cats and horses, it would revolve even faster. The advantage of such a feature is to enable the owner to be aware of the real value that the animal might be placing on the duration of any particular activity.

If worn by animals, the watches are preferably secured by a strap that is not easily chewed through.

FIG. 1

Apparatus for Training Cats to Use a Toilet Bowl

Encourage your cat to use the family toilet.

Cats are generally easier to look after than dogs as they do not need to be 'walked' and they tend to be cleaner when urinating or defecating. Despite this, their litter trays can be the source of highly unpleasant odours, and there is also the distasteful task of emptying and cleaning the tray. The current invention overcomes such problems by helping train a cat to relieve itself directly into a human toilet rather than a conventional litter tray.

Briefly described, the invention consists of a receptacle that can be attached to a toilet bowl. The device takes up approximately ⅙ of the open area of the toilet bowl so that the toilet can be used by both cats and humans without the receptacle being removed.

If, over time, progressively smaller amounts of cat litter are placed in the receptacle, the cat will gradually be trained to use the receptacle without the need for cat litter in it. At that point, the apparatus can be totally removed and the cat will relieve itself directly into the toilet bowl.

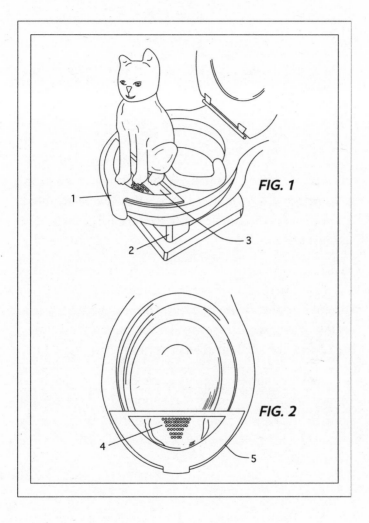

FIG. 1

1

2

3

FIG. 2

4

5

Automated Pet Petter

Provide your pet with the attention it needs when you are too busy.

As long as domestic pets have been in existence, they have had a special, dependent relationship with humans. This relationship has continued into modern times, but due to the hustle and bustle of modern life, the quality of the relationship has often declined.

One of the defining elements of this bond is the stroking, scratching and petting of an animal that can be achieved by means of a person's greater dexterity. Although this petting is essential to the relationship, it can frequently be neglected, undoubtedly to the annoyance and frustration of the animal involved.

To replicate this desired petting without the need for a person being present, the current invention uses a simulated hand-like element for pet contact. It simply needs the animal to be in the close vicinity of the device for all of the scratching, stroking and petting required to be carried out by the device itself.

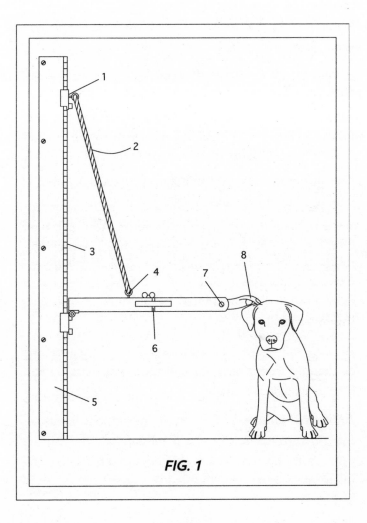

FIG. 1

Balanced Meal Tray

Let your pet enjoy a balanced meal with the rest of the family.

The current invention completely revolutionises how household pets are fed. Instead of providing the pet with one food item, the current invention provides two or more different food items separately. In this way, just as people do, the pet can enjoy different food items with a variety of tastes, smells and textures in a single sitting.

The innovative method consists of providing the pet with a main course and a treat, the treat being akin to the 'dessert' consumed by humans. To enable this, the invention provides a tray with two separate compartments. The main course is put into the first compartment and the treat is placed in the second. The tray is then presented to the animal and the animal consumes the contents.

In another design, the tray has three compartments, combining an enjoyable main course, a treat and a side dish in a single feeding.

Another advantage of the current invention is its ease of use. As with the rest of the family's meals, these meals can simply be removed from the freezer, placed in a microwave oven and then served.

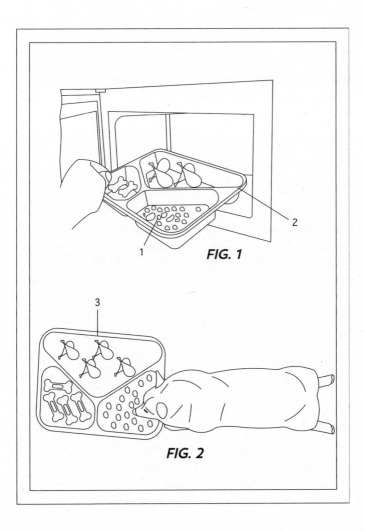

FIG. 1

FIG. 2

Cat Entertainment Box

Provide hours of fun for you and your cat.

The present invention is an enchanting item for cats which will also delight cat owners. The system comprises a four-sided cardboard box. The cat enters the box through a hole in the top which reads: 'Feline fantasy frolic.'

Each side of the box is decorated with an adorable sketch of one of four characters: a cowboy, 'Cowboy Cat', a circus weight-lifter, 'Strong Cat', a chic cat called 'Glamour Kitty' and a ballet dancer named 'Ballerina Kitty'.

A hole where each character's face would be enables the cat inside to poke its head out, and in this delightful way, the cat becomes the character on that side of the box. Enchanted cat owners can watch, and even take photographs, as their cat peeps out and pretends to be one of the different characters.

As all feline lovers know, cats are mesmerised by cardboard boxes and will play for hours and even go to sleep in them.

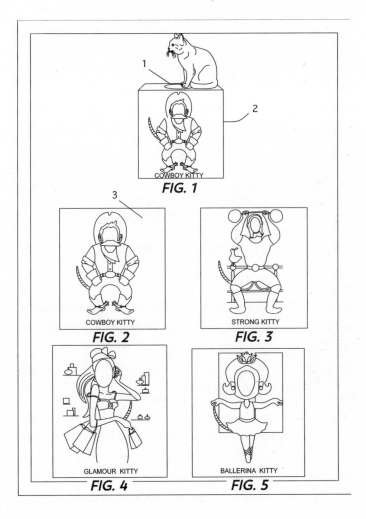

FIG. 1

FIG. 2
COWBOY KITTY

FIG. 3
STRONG KITTY

FIG. 4
GLAMOUR KITTY

FIG. 5
BALLERINA KITTY

Combined Aquarium and Cat Display Case

Display your cat alongside your tropical fish.

As everyone who owns both an aquarium and a cat is aware, cats are instinctively attracted to the fish swimming inside an aquarium. It is also well known that curiosity can get a cat into trouble. Cats have been known to leap on top of aquarium lids and use them as perches. Cats have been known to knock aquariums over. Some cats have even been known to fall into aquariums.

It is therefore an object of the present invention to overcome these problems for people who desire to own both a cat and an aquarium.

Located inside the invented aquarium is a smaller inner air chamber display case. This case is used to display a cat. The cat gains access to the case through an opening in the side of the aquarium and a special passageway connected to the display case.

By this innovative invention, people who own both a cat and an aquarium may thus increase their pleasure and enjoyment by viewing a combined aquatic and cat habitat which provides the illusion of their cat living underwater alongside their tropical fish and other aquatic creatures.

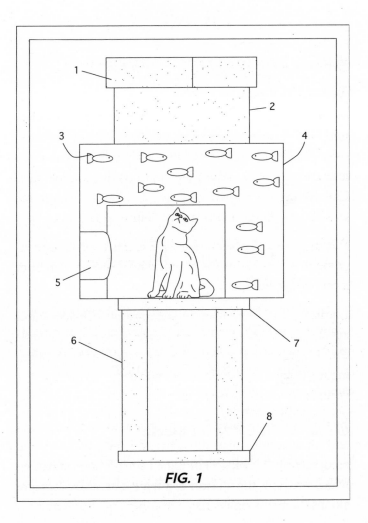

FIG. 1

Easily Emptied Petfood Can

Thoroughly empty your pet's food can using a drinking straw.

The current invention provides an original and unique device intended to overcome the common problems encountered when emptying petfood cans and other containers using a spoon or similar utensil.

At first glance, the container is similar to the typical cans used for dog food or the like; however, it has a number of unique aspects. First, the top lid is larger than the bottom lid, creating an angle in the can wall. Secondly, the bottom lid has a small vent hole which is covered by an air-tight seal.

To empty the container, the top lid is first removed with a conventional can opener. Next, the container is turned upside down and the vent seal tab is removed. A common drinking straw is then inserted into the vent hole and pressure is applied by blowing through it. This pressure, together with the force due to gravity, pushes the contents downwards out of the can.

Because of the slight angle in the walls of the can, the contents lose contact with the sides of the container and are thereby easily removed.

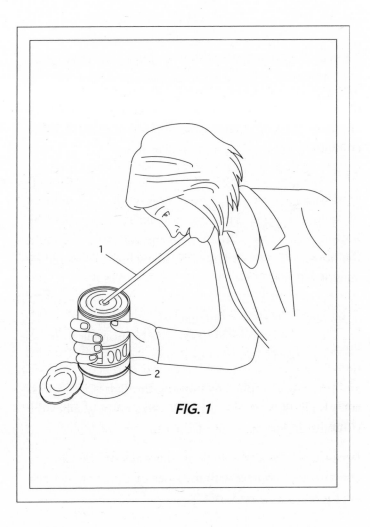

FIG. 1

Exercise Device for a Cat

Keep your cat fit, active and alert.

The present invention relates to an exercise device for a pet, and more particularly to an exercise device for a cat. Such exercise devices are currently almost non-existent.

The cat exercise device embodied in the present invention is composed of a base, a machine body, a drive mechanism, a lure, four loading members and four covers. The lure is an artificial mouse and is attached to the transmission belt such that it can be moved along with the transmission belt.

In operation, the cat requiring exercise is placed on one of the loading boards before the device is started. Once the cat is in place, the transmission belt is then actuated. This starts the mouse moving. As it circulates around the machine, the cat is lured to exercise by jumping from one loading board to another as it chases it.

The covers are there to conceal the mouse for a short period of time so as to promote the desire of the cat to hunt for prey.

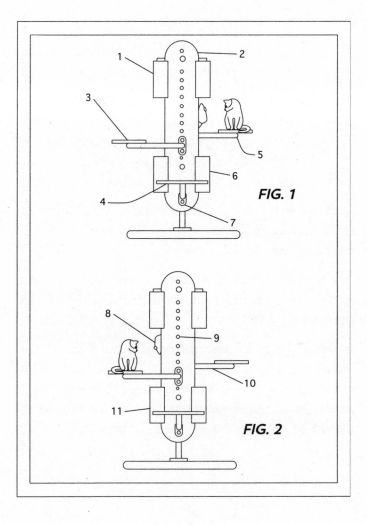

FIG. 1

FIG. 2

Pet Display Clothing

Show off your small pet with pride.

The current invention provides clothing for carrying and exhibiting small pets while worn by a person.

The clothing consists of a vest or belt with tubular passages for the pet which extend around the wearer's body. The outer walls of the pet passages are transparent so that when the clothing is worn, a pet moving along the passageways is visible to any interested onlooker.

Both the belt and vest can accommodate a variety of small animals, including mice, hamsters, rats, insects and possibly even snakes.

In alternative versions, the vest can have sleeves to form a jacket and also be of increased length to serve as an overcoat.

It should be noted that the corners and joins of the chambers are reinforced to increase strength and provide crush resistance. However, the wearer should still take care to guard against bumps and falls in order to reduce the risk of crushing the pets enclosed therein.

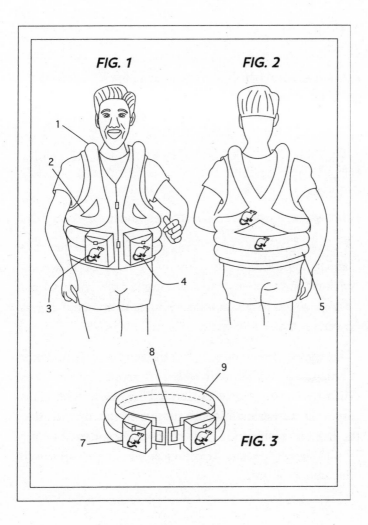

FIG. 1

FIG. 2

FIG. 3

Scary Dog Identification

Let passers-by easily know whether your dog might attack them.

The most reprehensible animal behaviour is violence and hostility towards other animals and people. Hostile and unsocialised pets can act aggressively and even attack passers-by and other animals with little warning. Also, with some animals, the usual signs of impending aggression, such as raised fur and flattened ears, many not be readily apparent. This means that those around the pet do not know whether it is safe to go near it.

A clear, distinct way of easily identifying whether it is safe to approach a pet is therefore needed. The present invention achieves this by providing a collar with a colour coded bandanna that features illuminating lights, warning signs and brightly coloured materials.

The use of red indicates that the animal is dangerous, orange that it is unpredictable and green that it is safe. An advisory message is positioned below the lights. This message further indicates to passers-by whether the dog is dangerous, unfriendly or friendly. Once alerted by the visual signals, passers-by can elect whether to approach the animal or not.

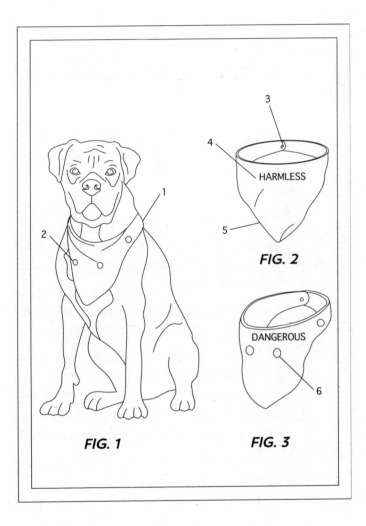

FIG. 1

FIG. 2

FIG. 3

HARMLESS

DANGEROUS

Small Animal Carrier

Pick up and carry your pet like a small suitcase.

In densely populated sections of major cities, high-rise apartments do not generally provide garden areas in which the owners of pets (specifically small dogs) can walk their animals outdoors. Instead, the owner must take their pet to a nearby park, frequently crossing busy vehicle intersections *en route*. Not infrequently, the weather may be inclement, necessitating that the pet, if small, be carried at least part of the way.

Such owners almost always soil their own clothing by picking up the pet and carrying it held against their bosom after, of course, it has already soiled its feet in wet or muddy areas of the park or walkway.

Accordingly, it is highly desirable to provide a device to enable pet owners to easily and conveniently pick up and carry their pet without the pet's soiled feet ever coming into contact with the owner's clothing. The present invention is such a device. It provides an apparatus so that a small pet can be easily carried, like a small suitcase, in a position such that its mouth and feet are prevented from reaching the person carrying it.

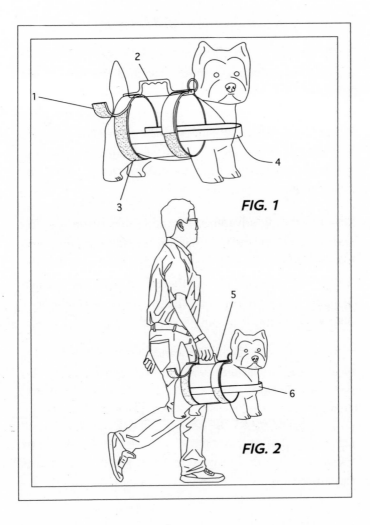

FIG. 1

FIG. 2

Vacuum Cleaner for Dogs

Vacuum clean your dog without frightening it.

In the past, vacuum cleaners have been devised for use on dogs for cleaning clipped hair and the like. These have not met with success, due to the fear instilled in a dog at the sight of a vacuum cleaner and by the sound of its very loud noise. An object of the current invention is therefore to overcome these disadvantages by providing a vacuum cleaner enclosed completely within the body of a toy dog.

An outstanding advantage of the combination toy dog and vacuum cleaner is that the vacuum cleaner and blower are not only completely hidden from view, but the frightening noise emanating therefrom is greatly muffled by being inside the toy dog. This enables a person to vacuum a dog without causing fear to the dog.

Additionally, the vacuum cleaner also functions as a blower, and air issuing from the tail can be heated to serve as a dryer.

After completion of the vacuum cleaning, the dog's tail can be automatically retracted into its original position by the use of any well-known retractable reel device, such as those used for air hoses in service stations.

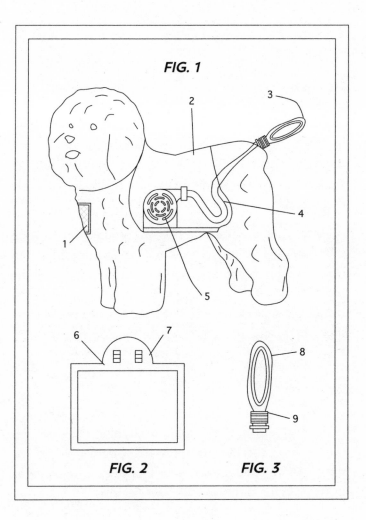

FIG. 1

FIG. 2

FIG. 3

For Parents

Baby-patting Machine

Take the effort out of putting your child to sleep.

It is generally known that it can be difficult to get a baby to fall asleep. In such instances, it is not uncommon for a parent to try patting a baby to sleep by repeated pats upon the hind parts thereof. This can be a time-consuming operation, particularly when the infant is unsettled and unlikely to fall asleep easily. It can be particularly displeasing if the child wakes during the night, thereby disturbing the parents' own sleep. This situation is in want of improvement. Accordingly, it is the principal object of the present invention to provide a machine which will pat a baby to sleep in place of the parent.

The device consists of a bracket with a motor which moves an arm with a soft pad in the form of a glove or mitten. When the motor is turned on, the arm moves backwards and forwards, giving periodic pats upon the rump or hind part of the baby. This simulates a pat from the hand of a person, thereby eliminating the necessity of a parent standing by and patting manually.

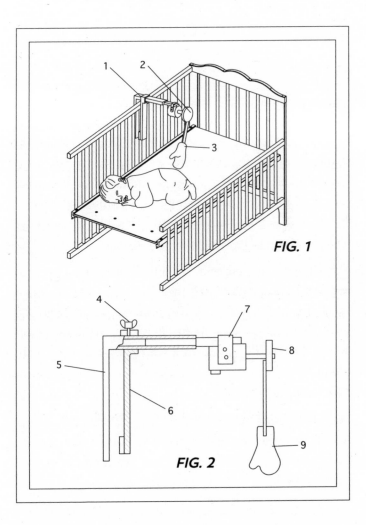

FIG. 1

FIG. 2

Paternal-bonding Bib with Breasts

Make paternal–infant bonding easier.

Maternal parents have traditionally taken on the major responsibility for the care and feeding of infants, however, more recently, paternal parents have assumed a greater share of that role. Studies have shown the positive influence of this bonding is enhanced if the male parent participates more significantly in nursing procedures. Few devices, however, are available to facilitate such nursing and its resultant bonding.

Accordingly, the current invention aims to improve bonding between child and male parent by anatomically simulating the nursing female. The device is given its quasi-anatomically true breast-like shape by the presence of suitable stuffing to form a mammary-shaped pouch which holds a container of liquid, such as milk, from which a nipple protrudes.

The device may be provided in any suitable contrasting colours, and the nipple area may be coloured with a rainbow pattern to attract and maintain the attention of the nursing child.

The device can also function as a tote bag for the father to carry nursing paraphernalia when required.

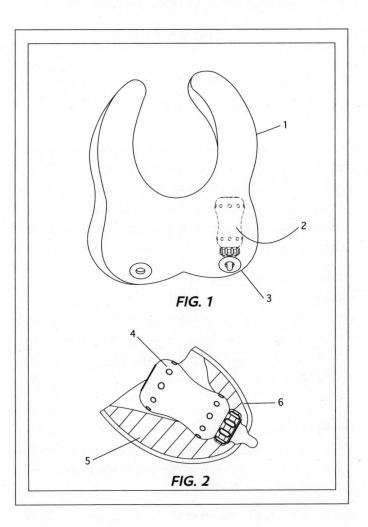

FIG. 1

FIG. 2

Combined Jacket and Nappy Bag

Wear everything your baby needs in one all-purpose garment.

Leaving the house with a small infant in tow generally requires a high level of organisation. Even if it is just for a short stroll in the park or a quick trip to the shops, parents and caregivers frequently find themselves in a high-stress situation when taking an infant outdoors. To reduce the stress, many parents often carry items such as nappies, wipes, creams, feeding bottles, cups, food, etc., in a nappy bag.

However, as many parents will testify, digging into the nappy bag while simultaneously consoling a screaming child is far from easy. Parents and caregivers also tend to carry their personal belongings, such as smartphones, sunglasses, keys, credit cards and the like, in a separate bag, adding to the difficulties.

The present invention addresses these problems by providing tech-savvy caregivers with a jacket, coat or vest that is especially designed with multiple pockets for infant care essentials such as nappies, feeding bottles and food, as well as their own personal items, thus functioning not only as a stylish garment for everyday wear, but also as a nappy bag and a handbag.

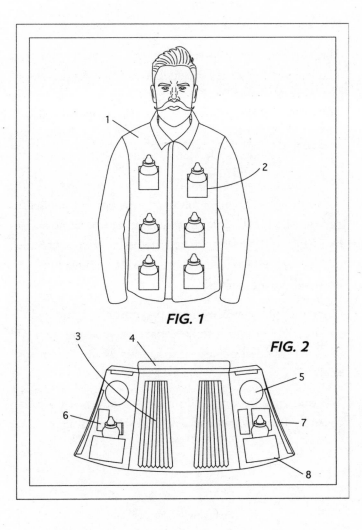

FIG. 1

FIG. 2

Infant-carrying Suit

Pick up your infant as easily as a small shopping bag.

The current invention relates to a suit for an infant or toddler which includes a special safety component to facilitate the quick and easy grabbing of the infant in situations of immediate or impending danger.

The suit covers the entire body of the infant, with a band around the front and handles on the back. The advantage of the device is that if the infant is in a position of danger, an accompanying adult can simply grab the handles and quickly lift the infant away. This can be done single-handedly and almost instantaneously, saving precious time compared to having to grasp a struggling child with both hands, for example.

Once the infant has been grasped and lifted, the band is of sufficient strength to be capable of holding the child in the suspended position for as long as the parent may wish.

The suit may be worn by infants of all ages, including those in the crawling, toddling and walking stages.

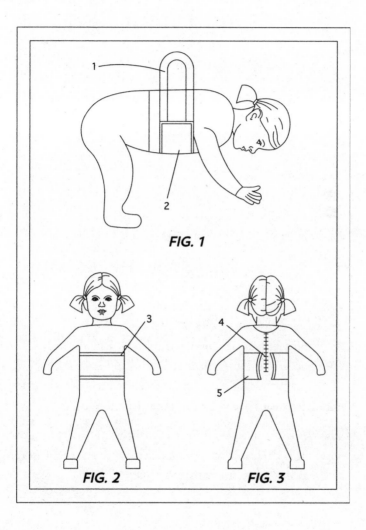

FIG. 1

FIG. 2

FIG. 3

Apparatus for Comforting an Infant

Let your child feel hugged and secure when you are too busy to comfort it.

It is well established that an infant is comforted by being embraced by a mother or other carer. It is also known that an infant finds a gentle touching motion particularly comforting. However, due to pressures on their time, a mother or other carer cannot always give their full attention to a child. A need therefore exists for a device which will effectively simulate a mother's embrace when she is unavailable.

To achieve this, the present invention utilises a body with a pair of arms which can be moved from an open to a closed position in order to embrace an infant. In use, the infant is placed against the body and the arms are wrapped around it. A mechanism is built into at least one of the arms to facilitate a movement which will simulate a mother's touch and thereby comfort the infant.

The device can be used in many different environments where it is desirable to soothe or reassure an infant, such as the home, nursery, car or the like.

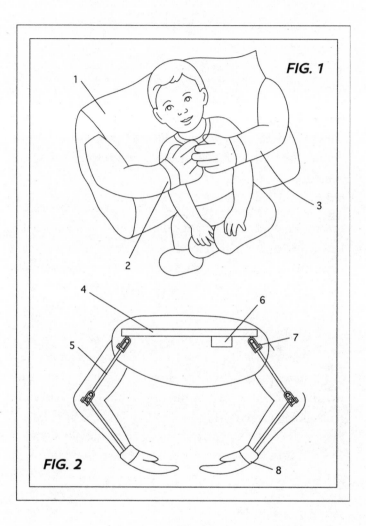

FIG. 1

1

2

3

FIG. 2

4

6

7

5

8

Non-slip Bedding

Make sure your child's bedding doesn't fall off by securing them to it with Velcro.

A very common problem for families with young children is that the children sometimes kick off their quilt, duvet or blanket while in bed. Another problem is that when children change their sleeping position or wiggle about, the quilt will also move and sometimes slip off their body.

In light of this, the inventor has devised the present invention, which will securely hold a duvet, quilt or blanket firmly over a child's body and prevent it from being removed or kicked off, either accidentally or intentionally.

To achieve this aim, a small sheet is fastened securely around the child. This sheet is provided with fasteners, such as the ones sold under the trademark of Velcro. The main quilt is also provided with complementary Velcro fasteners. The child can then be attached to the quilt by means of the Velcro, thus preventing the said quilt from being kicked off.

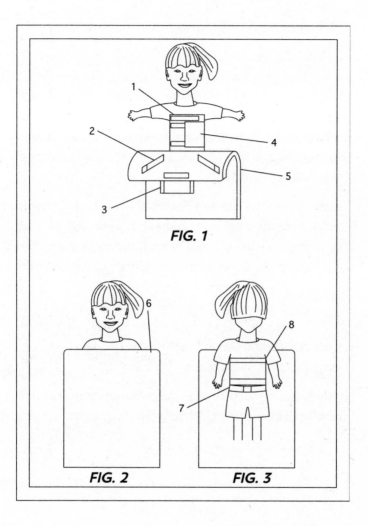

FIG. 1

FIG. 2

FIG. 3

Infant Crawler

Make learning to crawl fun for you and your child.

Devices for infants who are learning to crawl generally include a wheeled frame so designed that the child's hands and feet can contact the ground in order to move the crawler forward or backwards. However, such crawlers generally have the frame exposed, giving them a purely functional appearance with little play and entertainment value.

Accordingly, the present invention provides an infant crawler which can have many different shapes, such as animals, secured to the wheeled crawler frame. For example, if the cover is in the form of a turtle, by moving the crawler around in a room, the infant will give the appearance of a turtle moving about on the floor. This adds to the play value of the crawler and makes it more fun for the user of the crawler and also for people watching, such as parents or friends.

Other shapes, such as a ladybird, spider, butterfly, automobile, train, aeroplane, rocket ship, etc., can be provided.

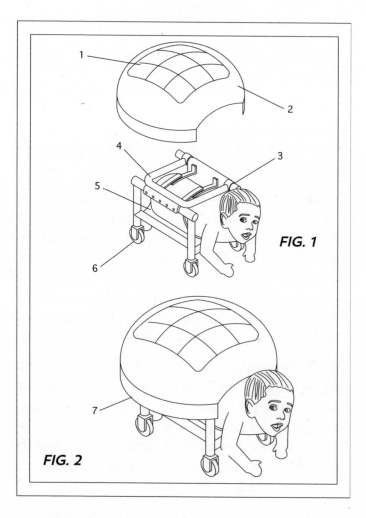

FIG. 1

FIG. 2

Interactive Toilet-training Apparatus

Attach your child's iPad to the toilet to help their toilet training.

Potty training young children is one of the greatest challenges a parent will face, and as many will testify, the unenviable business of enticing toddlers to use the toilet can be extremely frustrating.

With this in mind, the present invention is designed to assist parents in potty training their toddlers in a more engaging way. The apparatus consists of a support structure and a series of hook-and-loop straps, such as Velcro, designed to attach an interactive device to the inside of a toilet-seat lid. The device may be any device, such as an iPad Mini, a Kindle Fire, an etch-a-sketch or a pinball game, that will keep a child's interest.

In use, the apparatus is fixed to the interior side of the toilet seat lid and the chosen device is then attached to the apparatus. The child, who should be placed facing the cistern, can then play with the device while squatting on the toilet bowl. This eliminates the boredom and frustration both children and parents can experience with toilet training.

An additional benefit is the reduced cost of nappies due to quicker and more efficient toilet training.

FIG. 1

Combined Toilet Trainer and Toy Car

Make learning to use the toilet fun and exciting.

Over the years, there have been many different devices to assist with toilet training children. While some of these devices provide added interest to the child beyond that offered by the standard potty, they still do not provide the child with a really fun way of learning how to use the toilet. Furthermore, as soon as the child learns how to use the toilet properly, these devices become obsolete.

It would therefore be beneficial to have a device that was fun and exciting for a child to use when learning to use the toilet. It would be an additional benefit to have a device that activated a horn and lights when a child successfully deposited a waste product. It would be a further benefit to have a device that could be used as a toy car after the child had learned to use the toilet. The present invention is such a device.

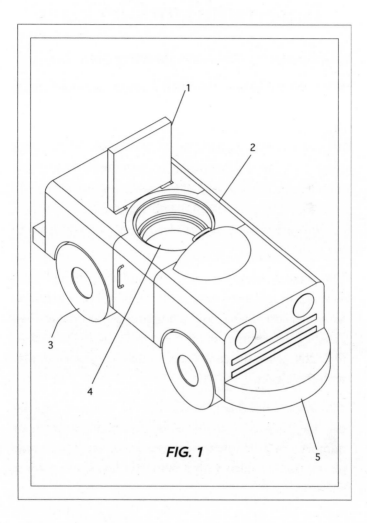

FIG. 1

Educational and Entertainment Necklace

Step out in style while educating and entertaining your child.

Entertaining children, especially when busy with other things, is a continual challenge for most parents. This is especially true in the case of very young children, who often demand to be held for long periods. To distract a child, a parent will often put an entertaining object into their hands. However, in almost every instance, the child can be guaranteed to drop the object on the floor, requiring the parent to pick it up, clean it and then return it to the child.

The present invention solves this problem by providing a device in the form of a cord or chain, like a necklace, to which interchangeable entertaining objects are then attached. In use, the device is secured around the neck of the adult, while the child is held within easy reach to play with the object.

The interchangeable entertainment objects may be educational, such as laminated cards with pictures, letters or numbers, or primarily for entertainment, such as flexible plastic objects filled with a colourful liquid to promote chewing.

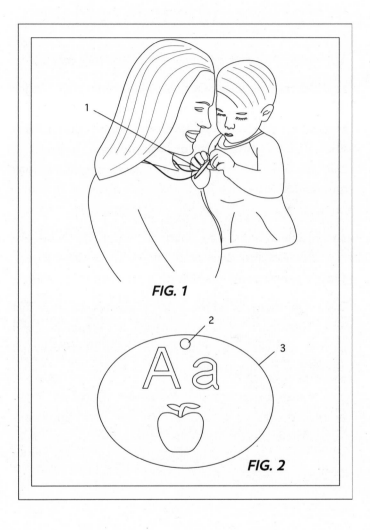

FIG. 1

FIG. 2

Wearable Child-activity Device

Effortlessly turn yourself into a walking activity centre.

It is generally agreed that providing a child with a wide range of physical, visual and audio stimuli during the early stages of their development can have lifelong benefits. Various activity centres have therefore been developed in order to encourage a growing child's physical and mental capacities.

Typically, these activity centres include a range of toys specifically designed to sharpen the child's hand–eye coordination and their familiarity with colours, letters, shapes, numbers and the like.

Although many of these activity centres are portable and easily attach to common items such as prams, child carriers and car-seats, none has yet been devised which attaches to the appendages of the caregiver.

The present invention is such a device. Moreover, it can be placed on two appendages of the adult simultaneously. Alternatively, due to the elastic nature of the material, it could be placed upon one or both of the child's appendages if desired.

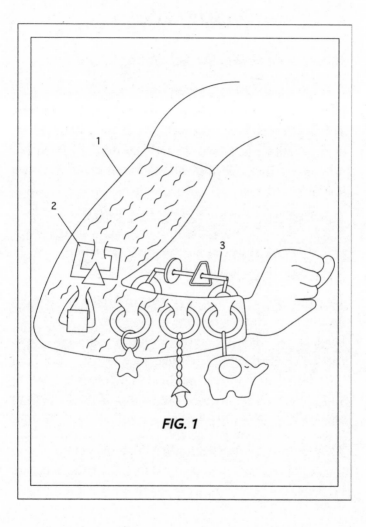

FIG. 1

Figurine for Displaying Baby Teeth

Never lose your baby's first teeth or hair.

It is not uncommon for parents to keep their children's baby teeth, hair and other items, such as clothes or toys, as a remembrance of their children's early years. However, at times, the baby teeth and hair may be placed in a box or envelope which is then lost or misplaced. If not lost, it is often the case that the items are very seldom, if at all, taken out and viewed. The situation can be even worse with the young child's clothing, which is often stored in an old box in an attic or cellar, where it might deteriorate or be accidentally thrown away.

To avoid such a sad and sorry state of affairs, the present invention provides a figurine in the form of a doll. The head of said doll has an open mouth with a groove in the lower portion thereof. The human baby teeth can be placed in the groove and held in position by glue. In addition, a hole is provided on the top of the doll's head where human baby hair can be placed.

The doll may be dressed in the child's own clothing and then put in a prominent place so that all of the items are on display for everyone to see at all times.

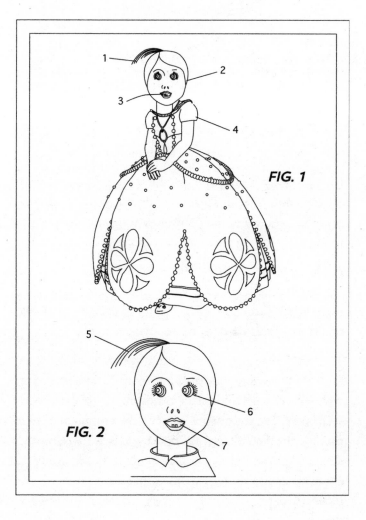

FIG. 1

FIG. 2

Car-trip Barrier Device

Ease the strain of long family car trips by separating your children.

Family car trips are a well-known high-stress operation. This is because children instinctively struggle with each other when sitting close together on a seat. They struggle over toys, territory, touching, food and views. Such tensions can rapidly escalate to alarmingly high levels.

In the past, parents have used a variety of methods to reduce the level of conflict and animosity. They have provided entertainment, shouted, resorted to threats or force, or simply decided to stay at home.

Each of these methods has major limitations. Providing entertainment requires time and effort. Shouting or using force can cause bad feelings and ruin the outing, and staying at home can be annoying.

To solve this problem more effectively, the present invention provides a flexible barrier to separate travelling children. When children or other individuals sit on the attached sheet, the barrier becomes semi-rigid and therefore blocks the children from struggling with each other. This makes the journey considerably less stressful for all concerned, and thereby helps to maintain good will and positive feelings between family members throughout the journey.

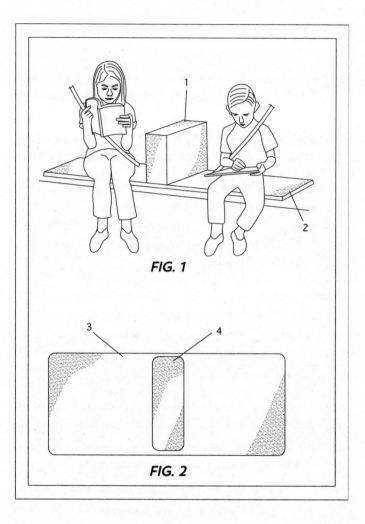

FIG. 1

FIG. 2

Motivational Sack

Replace the 'Naughty Step' with the 'Naughty Sack'.

As all good parents know, one of the main goals of parenting is to raise children in a way that promotes both good behaviour and healthy emotional growth in the child. It is also the responsibility of the caregiver to correct bad behaviour in a way that teaches the child a lesson while also encouraging an appreciation of the rules.

The present invention achieves these aims by providing a motivational and behavioural modification system wherein a child's 'lovey' (favourite item, toy, blanket, game, etc.) is placed in a 'Naughty Sack' every time the child breaks the house rules.

The Naughty Sack has negative indicia displayed on the exterior surface, which acts as a visual warning in order to encourage better behaviour. The negative indicia could be any image, graphic or text that sends a clear and unequivocal message to a child that the sack is 'BAD'.

Fig. 1 depicts a caregiver placing a 'lovey' in the motivational and behavioural modification system sack. The confiscated 'lovey' may be returned to the child after a prescribed period of good behaviour.

The system includes an illustrative instructional pamphlet for the caregiver on its correct use.

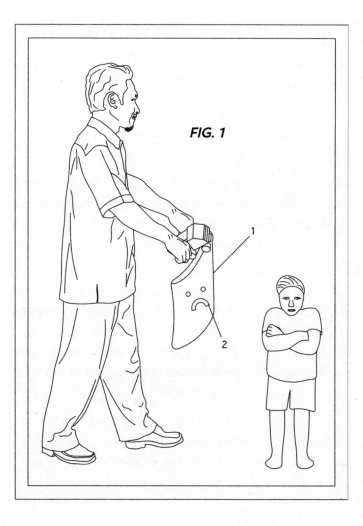

FIG. 1

1

2

Healthy Choices Reward Kit

Help your problematic child say no to drink, drugs and violence.

This invention relates to an educational doll and kit which help a child make healthy choices and not do bad things to other people every day of the week.

To use the kit, the child takes the doll and labels its hat with the label for that day of the week. For example, if it is Wednesday, the label 'WED' is put on the hat. Then the hat is put on the doll's head.

The chart of attachable stickers is then used. For example, if the child has not been violent that day, the sign 'I WASN'T VIOLENT TODAY' is placed on one arm of the doll. If the child did not drink alcohol, the sign 'I DID NOT DRINK ALCOHOL' is placed on the doll's other arm. If the child did not use drugs for fun, the sign 'I DID NOT USE DRUGS TODAY' is placed upon the leg of the doll.

The process is repeated for each day of the week, reinforcing the healthy choices the child makes each day.

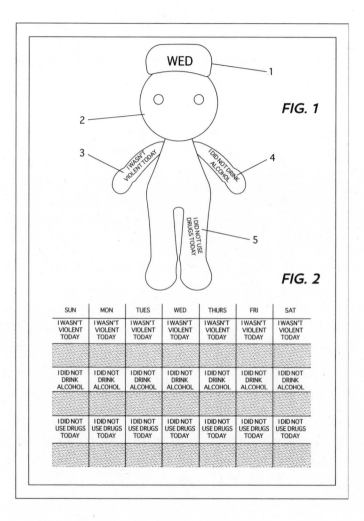

WED — 1

2

3 — I WASN'T VIOLENT TODAY

4 — I DID NOT DRINK ALCOHOL

5 — I DID NOT USE DRUGS TODAY

FIG. 1

FIG. 2

SUN	MON	TUES	WED	THURS	FRI	SAT
I WASN'T VIOLENT TODAY	I WASN'T VIOLENT TODAY	I WASN'T VIOLENT TODAY	I WASN'T VIOLENT TODAY	I WASN'T VIOLENT TODAY	I WASN'T VIOLENT TODAY	I WASN'T VIOLENT TODAY
I DID NOT DRINK ALCOHOL	I DID NOT DRINK ALCOHOL	I DID NOT DRINK ALCOHOL	I DID NOT DRINK ALCOHOL	I DID NOT DRINK ALCOHOL	I DID NOT DRINK ALCOHOL	I DID NOT DRINK ALCOHOL
I DID NOT USE DRUGS TODAY	I DID NOT USE DRUGS TODAY	I DID NOT USE DRUGS TODAY	I DID NOT USE DRUGS TODAY	I DID NOT USE DRUGS TODAY	I DID NOT USE DRUGS TODAY	I DID NOT USE DRUGS TODAY

For the Bathroom

Toilet Bowl Protector

Protect the rim and exterior of the toilet from your inability to urinate accurately.

The problem of urine collecting on the exterior of a toilet bowl has existed as long as men have used toilet bowls to urinate in. The problem is caused because a typical male usually urinates from a standing position. As a result, incidental sprays of urine frequently miss the toilet bowl and slowly trickle down the exterior of the toilet. Over time, this leads to an unsanitary situation and a distinctly unpleasant odour. Regular cleaning to prevent such an unpleasant build-up is a disagreeable task.

To overcome this problem, the present invention provides a protective toilet skirt made of plastic sheeting. This is attached to the toilet rim by a plurality of specially designed clips, or Velcro if preferred. When secured in place, the plastic sheet acts as a shroud to protect the exterior of the toilet bowl and other vulnerable areas, such as the toilet seat hinges and toilet rim, from sprays of urine. In extending outwardly from the base of the toilet, it also functions as a floor mat for extra protection.

As it is easily detachable and washable, it may be used and reused.

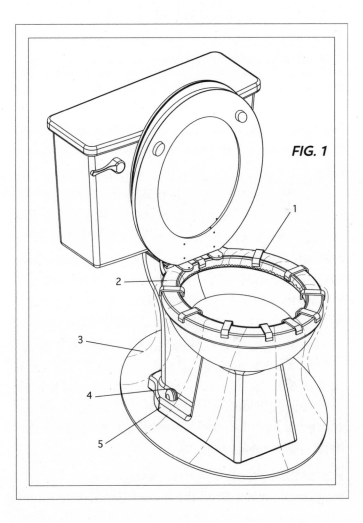

FIG. 1

Apparatus for Instruction in Toilet Use

Get 'real-time' reminders on how to use a toilet properly.

Learning about correct toilet use is a critical stage in every child's development. Lowering the toilet seat and lid, flushing the toilet and properly washing the hands afterwards are all key steps which need to be learned.

However, given the number of letters to advice columnists on this subject, it is clear a significant number of grown men never successfully complete their toilet training. While most men have successfully learned to regularly flush the toilet, many fail to lower the toilet seat and lid and wash their hands properly afterwards. The current device aims to remedy this.

A toilet use detector unit attached to the flush handle is used to monitor toilet activity and is configured to provide an audible message giving the person instructions on the next appropriate step in toilet use and hygiene. For example, if the flush handle is used, an audio instruction reminding the user to lower the seat and wash their hands will sound.

For added effectiveness, the message can be recorded by a female member of the household, making it more likely that the husband or boyfriend will heed the advice.

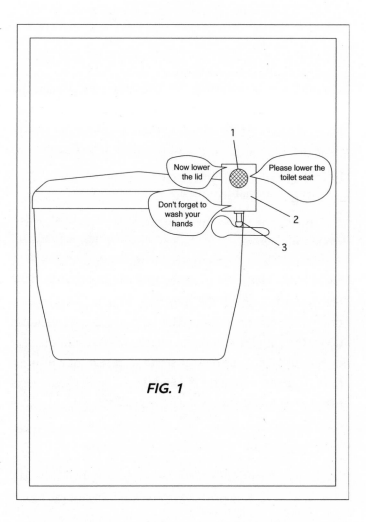

FIG. 1

Anti-constipation Device

Ease constipation strain by hanging over the toilet.

People suffering from constipation may sense an urgent need to defecate, but when they sit on the toilet they are not able to expel any faecal matter. In such a situation, they may consequently push and strain too much, causing haemorrhoids. Many people experiment with changes to their diet and over-the-counter remedies for relief. The present invention provides an innovative alternative to such conventional treatments.

The invention works by straightening a person's torso while the person 'semi-hangs' from a bar, similar to a chin-up bar, stationed above the toilet. The act of 'semi-hanging' causes a change in the toilet user's posture and unfolds part of the intestine. This induces a position which is more conducive to moving a faecal column down through the colon and rectum. As a result, the user does not have to push and strain as much, nor do they have to spend as much time endeavouring to expel faecal material as might otherwise be necessary.

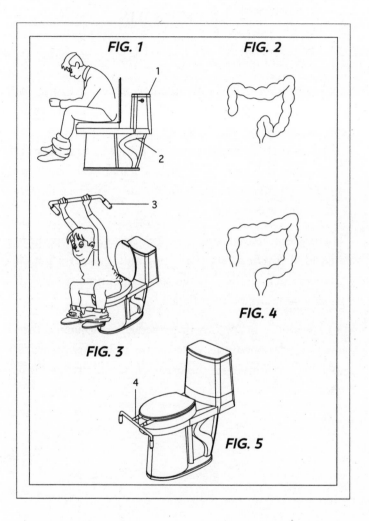

FIG. 1

FIG. 2

FIG. 3

FIG. 4

FIG. 5

Forehead Support Apparatus

Rest your head while urinating.

As many men will testify, needing to use a urinal and stand in an upright position when feeling tired and weary can seem excessively difficult and burdensome. Currently, in such circumstances, the only way to relieve the strains and tensions of an incapacitated body is to rest one's forehead on the wall directly above the urinal. This can be uncomfortable.

The present invention is therefore directed to providing a padded forehead support apparatus so that a person can rest their forehead when using the urinal for a more comfortable overall experience.

An alternative embodiment provides an elongated forehead support apparatus which spans a plurality of urinals so as to accommodate the users of all the urinals in the line.

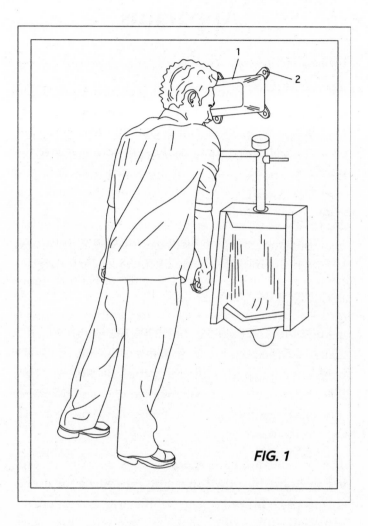

FIG. 1

Night-time Urinal Apparatus

Urinate into the toilet bowl even when your penis is pointing upwards in the dark.

No one enjoys cleaning a bathroom that has been used by men, especially the area around the toilet. This is because when males urinate, they frequently miss the inside of the toilet bowl and hit the toilet seat or the outside of the toilet instead.

This problem is particularly acute at night when a boy or a man using the bathroom may be too sleepy to pay proper attention and miss the inside of the toilet completely.

Also, when men wake at night to urinate, their penises are often erect or semi-erect. As the penis is then pointing upwards and the toilet bowl is positioned below, this makes it difficult to precisely deliver urine into the toilet bowl. Trying to push your erect penis down to aim at the toilet bowl can hurt, as every man knows, and no matter how much one tries, it can be impossible to prevent spillage around the toilet.

The present invention addresses this problem by providing a portable urinal apparatus into which the user is able to place his erect penis. In this way, he can urinate without the worry of accurately aiming at the toilet bowl, thus avoiding any splashing and spilling.

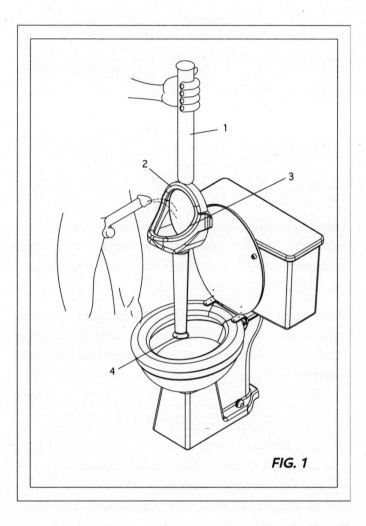

FIG. 1

Portable Vibrating Bidet

Discreetly pleasure yourself when you are on the toilet.

Toilet paper, which is a soft tissue-type paper, is generally the preferred method in many Western countries of cleaning oneself after defecation, urination or other bodily functions.

A different method of personal post-toilet cleaning is to use water, as with a bidet. (A bidet is a device which usually takes the form of a low-mounted sink which is specially designed for washing the genitals, buttocks and anus.) It is generally known that washing with water is more effective at removing deposits and waste residues from one's skin than rubbing with paper. Bidets, however, are still not common in many countries.

It is therefore an object of the present invention to provide a portable bidet which can be used in situations where a bidet is not available and which overcomes the limitations of prior designs.

Unlike previous portable bidets, the current invention provides an integrated dryer for the buttocks and genitals and includes a built-in heater so that the temperature of the cleaning solution may be regulated. It is also designed to be easily concealed and carried, and additionally includes a vibrator that may be utilised as a pleasure device if desired.

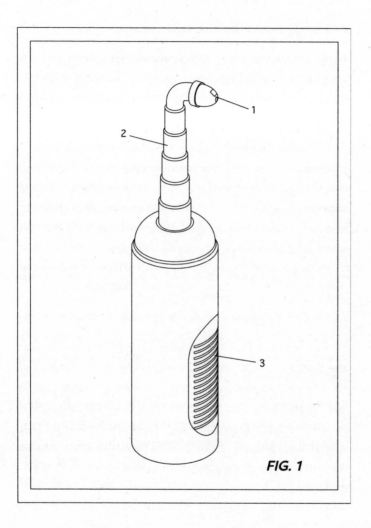

FIG. 1

Stay-dry Toilet Seat

Use public toilets with ease and assurance.

Using a toilet in a public rest-room facility can often be a particularly disagreeable experience due to the dirty or wet condition of the toilet seat.

When faced with such a prospect, a user is most likely to resort to slightly bending their knees in an attempt to maintain a hovering position over the toilet bowl without actually sitting on the toilet seat. Sustaining this mid-air position can, however, be uncomfortable, exhausting and precarious. It may also cause more splashes on the toilet seat if the user sways and wobbles.

It is therefore clearly desirable to have an accessory that enables the user to comfortably maintain a semi-standing or hovering position when using public toilet facilities.

To accomplish this, the current invention provides two thigh support pads carefully angled to facilitate the user maintaining a semi-squatting position over the toilet bowl while urinating or defecating. The pads are permanently connected to the toilet by a hard rubber frame which has enough 'give' to comfortably accommodate users of different heights and weights.

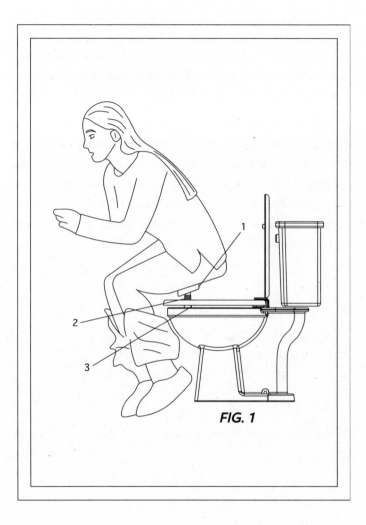

FIG. 1

Table for a Toilet

Read a book, enjoy a cigarette or make a shopping list on the toilet.

The toilet can more correctly be called 'the throne' if certain provisions beyond the conventional support and flush functions are met.

A reliable supply of toilet paper needs to be provided in a favourable location, not in a place which is difficult or impossible to reach. A range of reading material is also welcome, and for some, cigarettes, matches and an ashtray will be needed. Others may prefer to write notes, such as shopping lists, so a pencil, notepaper and backing for the notepaper will be required. All the aforesaid items should be provided with easy access and the least possibility of dropping or slipping.

The current invention does all this and is more convenient than any known system like it. The 'table' top is of a proper size and is helpfully sloped to hold a magazine. The system itself is easy to load with toilet paper, and as the contents are kept above ground, it allows for good visual inspection. There are also no accessible pockets to attract vermin.

In summary, it is durable, sturdy, handsome in appearance and large enough to be seen and not tripped over.

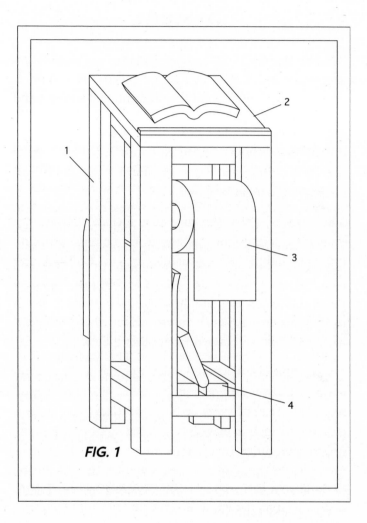

FIG. 1

Toilet Noise Recording System

Monitor your health by recording the noises you make on the toilet.

The current invention provides a system for recording and reporting toilet noise data. In use, one or more microphones and a connected controller allow for the detection, storage and transmission of the noise produced on the toilet by a user.

The types of noise recorded may include, but are not limited to, sounds such as bowel movements, urination duration and frequency, flatulence, etc. Collated user data may then be sent to the user's doctor by means of internet transmission, or may be stored in order to assess or monitor changes in the user's health when compared to related noises obtained from other toilets.

For example, a constipated toilet user may make sounds, such as groaning or grunting noises, that match pre-recorded noises of other constipated toilet users. Constipation information may also be based on bowel plopping noises and other bowel movement noise. In another example, the duration of urination noise and the related harmonics may match a predetermined urinary condition or bladder infection.

The system also allows for all recorded toilet noises to be linked to an individual user account and profile.

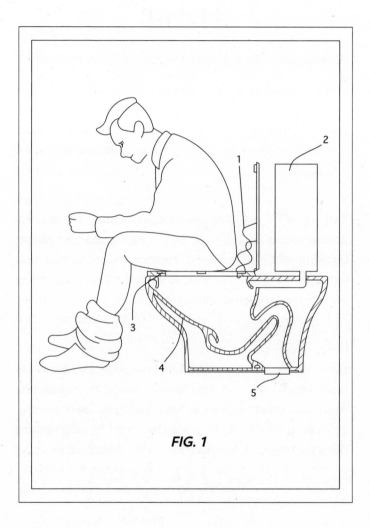

FIG. 1

Toilet Seat-lowering Device

Never forget to lower the toilet seat again.

Women generally appreciate the rare breed of men who remember to lower the toilet seat and lid after using the toilet. Constantly having to remind male family members to do this puts an unnecessary strain on family relationships.

Young male users can be especially troublesome in their failure to recognise the practical and aesthetic value of a lowered toilet lid and seat, while older males may simply be thoughtless or stubborn. Either that, or the male toilet user may simply have other things on his mind when responding to the call of nature.

Usually most males, both young and old, at least remember to flush the toilet after use. The present invention takes advantage of this by blocking access to the toilet flush handle when the lid or seat is up. To engage the flush, the toilet user must first lower the seat and lid. The device also includes a warning graphic, emphasising the advantages of lowering the toilet seat and lid.

Over time, the user will find it will become a habit to lower the toilet seat and lid, finally bringing an end to disagreeable confrontations between the females of the house and the offending males.

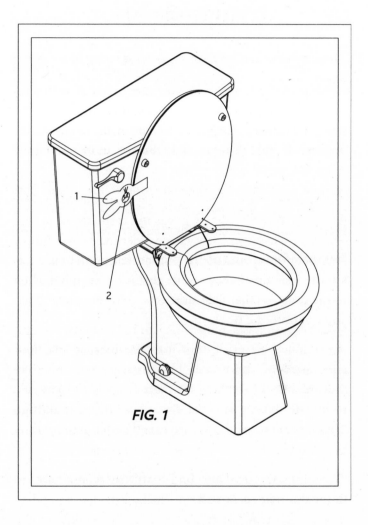

FIG. 1

'Please Stand Here' Urination Spot

Avoid urinating on the floor by always standing in the correct place.

Public lavatories for men are usually furnished with wall-mounted urinals. However, it is known that a significant number of men using such facilities tend to stand away from the urinal while urinating. As a result, urine frequently drips onto the floor in front of the urinal. This quickly results in unpleasant and unhygienic conditions that can only be mitigated by frequent cleaning.

It is therefore an aim of the present invention to provide a means to encourage men to position themselves close to the urinal so that their urine goes into the urinal basin and not onto the floor.

Accordingly, a pair of foot mats are fixed to the floor either side of the urinal. The mats are positioned so that the person standing on them is properly stationed to urinate directly into the urinal basin. The inscription 'Please stand here' clearly instructs the user in the correct positioning.

The invention could also be beneficially adapted for use in the domestic bathroom.

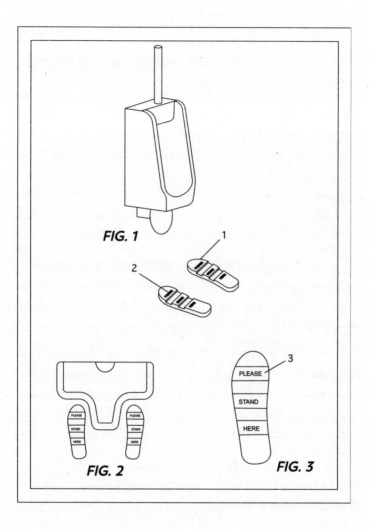

FIG. 1

1

2

FIG. 2

PLEASE
STAND
HERE

PLEASE
STAND
HERE

3

PLEASE

STAND

HERE

FIG. 3

Toilet-squatting Aid

Easily take up the squatting position on a conventional toilet.

Since man first roamed the Earth he has used the 'squatting position' to discharge his bodily functions, and young children of every creed and culture naturally adopt this position when they need to relieve themselves. Today's seat-like toilet, however, invented in western Europe several hundred years ago, rapidly became a way of highlighting the difference between 'civilised man' and more 'primitive' peoples.

This departure from the natural squatting position to a more 'civilised' seated position is considered by many health professionals to be a key factor in a number of health problems affecting westernised countries. Common ailments such as constipation, hernias and haemorrhoids are just some of the problems linked to the use of the seat-like toilet.

The primary aim of the current invention is therefore to provide a portable toilet accessory that enables a person to use the natural squatting position on a conventional modern toilet.

The device fits easily over all standard toilets. The anti-skid feet, located at the base of each support, help prevent the unit from slipping on the bathroom floor while in use.

FIG. 1

Toilet Waste Chopper Kit

Easily chop up any colossal stools to prevent your toilet from clogging.

Conventional toilets may not flush properly if the waste therein (e.g., stool) is too considerable in size. Such over-sized waste can clog the toilet and cause it to overflow. The present invention therefore provides a solution in the form of a chopper kit that can be used to facilitate the flushing of bulky waste down the toilet and prevent toilet clogging.

The invention features a container and a chopping blade which is attached to the end of a shaft. Before flushing the toilet, a person who has deposited an excessively large stool can use the blade to chop the waste into smaller pieces to facilitate flushing and avoid toilet blocking.

The blade can be of any shape and size that permits it to move freely in the space at the bottom of the toilet bowl where stools generally settle. In one version, the lower portion of the blade has a heart shape with a rounded tip so that it can more effectively chop up a stool lying at the bottom of the conventionally rounded toilet bowl.

Various modifications of the invention, in addition to those described, will be apparent to those skilled in the art.

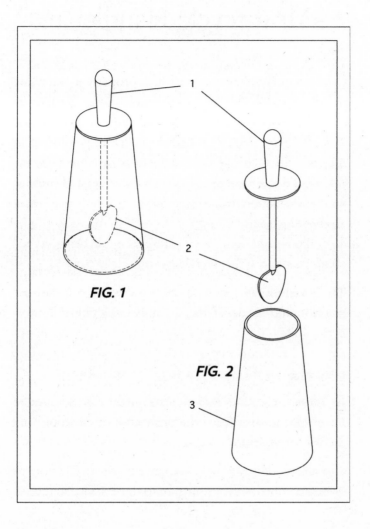

FIG. 1

FIG. 2

1

2

3

Urinal Controlled with a Motorcycle Handlebar

Pretend you are riding a motorcycle when using the urinal.

The present invention pertains to male urinals in general, and more specifically to a replica motorcycle handlebar that operates the flush function of the urinal.

The invention includes a connector that runs from the throttle on the handlebar to the flush valve of the urinal, so that the action of carrying out a throttling movement flushes the urinal. The throttling action may also trigger an audio recording of a motorcycle engine being revved so as to enhance the overall motorcycle experience.

The invention also includes mirrors that match the size, position and shape of those found on a typical motorcycle, adding to the overall motorcycle effect. Additionally, the invention may also include a button that when pressed emits the sound of a motorcycle horn.

In these and other regards, the urinal of the present invention departs from the conventional concepts and designs of previous urinals.

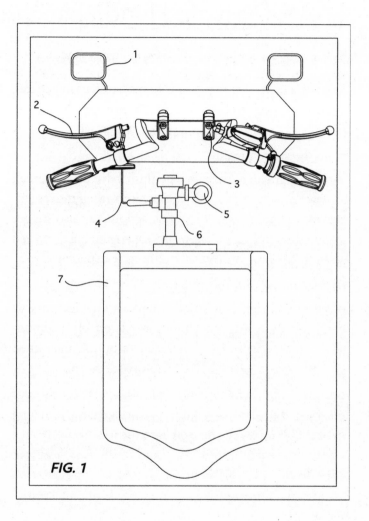

FIG. 1

Urination Kneeling Aid

Make less of a mess by kneeling down to urinate.

The established practice of human males standing up to urinate into a conventional toilet inevitably results in urine and toilet water splashing onto the exterior of the toilet bowl as well as the floor and wall around the toilet. This soon leads to an objectionable smell emanating from the soiled area.

The present invention seeks to alleviate this problem by utilising the fact that the quantity of liquid splashed from the toilet bowl is a function of the height from which the urine falls. It therefore follows that if the distance the urine needs to fall is lowered, there will be less splashing of the area around the toilet.

The invention thus allows the male user to urinate from a lower height by enabling him to kneel down next to the toilet bowl. When the need to urinate arises, he simply has to approach the toilet bowl, clench the handgrip with one hand and then lower himself down so that one knee rests on the platform. Thus positioned, the urinating male is able to relieve himself into the toilet bowl from a height only slightly above the bowl itself. This results in considerably less splashing and fouling of the surrounding area than previously known methods of urination.

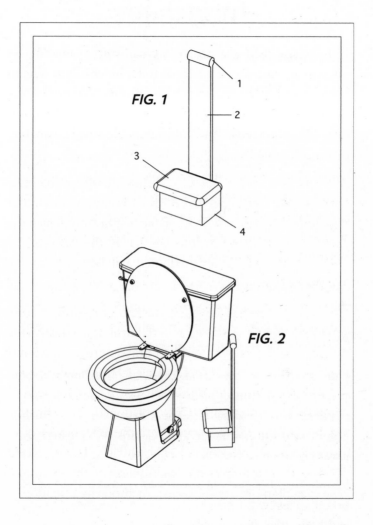

FIG. 1

1
2
3
4

FIG. 2

Toilet Seat Volatile Gas Incinerator

Banish obnoxious toilet odours by burning them.

When defecating, the human body generates odours that are considered by many to be objectionable.

One solution has been to mask the gases using suitable masking agents. These cover the initial obnoxious smell with another smell that is hopefully less noxious and may even be pleasing. They carry such fresh-sounding names as Autumn Leaves, Rose Fragrance and the like in an attempt to convince the user that the new odour is in fact as fresh as the outdoors and not at all obnoxious.

Another solution has been to discharge the gases into the surrounding atmosphere, but this does not eliminate the problem given the close proximity and high-density lifestyle of the modern urban dweller.

Until the present invention, there has been no suitable system to solve the problem. In use, a vacuum pump connected to a hollow toilet seat draws air through the hollow seat into an incinerating device of tungsten grid wires and then passes the air outside.

Simply put, the present invention removes the volatile bodily gases by burning them, thereby removing the gas and hence the objectionable odour.

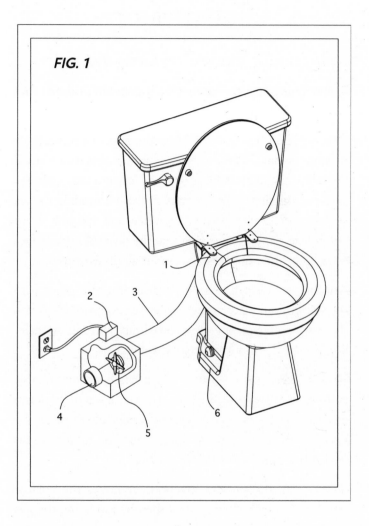

FIG. 1

For Sex

Decorative Penile Wrap

Heighten your partner's arousal by decorating your penis.

Prior to actual sexual intercourse, humans are typically known to embark on acts of foreplay, with the aim of intensifying the sexual arousal of both partners. Along with direct physical contact, they are also known to take advantage of various devices to heighten such arousal.

The present invention provides such a device in the form of a decorative wrap which is placed around a human penis to increase the sexual arousal of both partners prior to intercourse or other related sexual activity.

Said decorative wrap may be secured to the penis by a variety of means, including an elastic band, tie straps, Velcro-type fasteners or other suitable fasteners mounted at one or more locations along the side of the sheath. A decorative indicia is then attached to the exterior of the sheath to provide a stimulating appearance. The decoration can take the form of an inanimate object, such as a snowman, golf bag, ghost, wizard or reaper, or may be in the form of an animate object, such as an animal. The choice would depend upon the desires of the couple concerned.

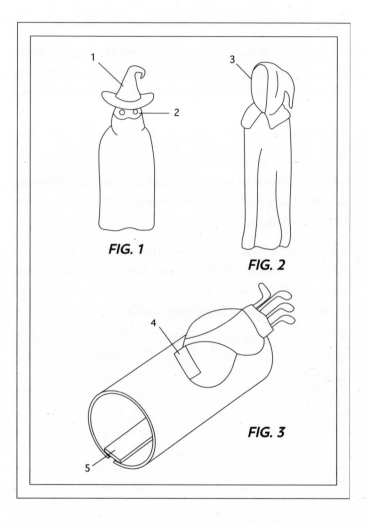

FIG. 1

FIG. 2

FIG. 3

Textured Tongues

Add a little variety to your experience of cunnilingus.

Throughout human history, individuals, especially women, have frequently expressed how difficult it is to orgasm through the act of sexual intercourse. As a result of such frustrations, many marriages and relationships have been known to fail.

The present invention addresses this problem by providing a sexual aid to significantly elevate the experience of oral clitoral stimulation and help a woman reach the ultimate climax.

The device is made out of a thin, stretchy material and fits on an individual's tongue like a sleeve. Imagine a condom for the tongue. It is secured to the tongue by an elasticated band and reinforced strap.

The device has a range of textures with soft protuberances on the surface to heighten stimulation. A perforated version permits the exposure of parts of the user's tongue. Each version of the device will be available in a variety of sizes to accommodate different tongue sizes.

It will be apparent to those skilled in the art that the textured tongue is not limited to clitoral stimulation and may also be used on the penis, nipples, testicles, anus and other body parts.

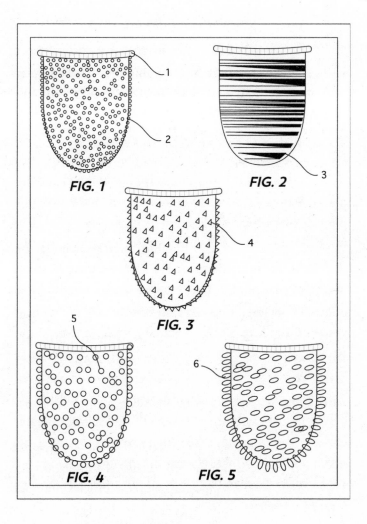

FIG. 1

FIG. 2

FIG. 3

FIG. 4

FIG. 5

Intimate Pool Lounger

Enjoy oral sex on an inflatable lounger without the risk of capsizing.

It is generally known that swimming pools host a range of different activities beyond traditional swimming, ranging from water games to sex.

To aid such non-swimming activities, floating loungers are commonly used, however, traditionally such devices have not been designed for activities involving two people. In many instances, the combined weight of two people will sink or capsize the lounger, and even if it does not sink or capsize, it can be extremely difficult for two people to find a comfortable position.

The present invention solves this problem by providing a conveniently sized opening at approximately the midway point of the lounger. The opening enables two people to have sex, and especially oral sex, without sinking or capsizing.

In use, the person receiving cunnilingus or fellatio lies on the lounger with their legs open and their genital area close to the opening, while the person performing has their head protruding through the opening. This person could either be standing on the bottom of the pool or floating.

It should be apparent that the opening is not restricted to oral sex and may also provide access for intimate massage.

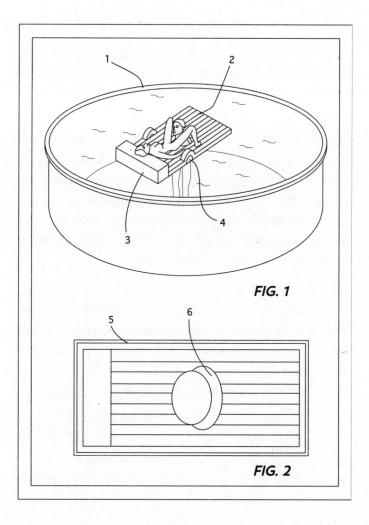

FIG. 1

FIG. 2

Artificial Love Handles

Ensure you or your partner can always get a good grip.

In recent years, along with the so-called sexual revolution, men and women have become more open with one another about sex, its pleasures and what each person desires the other to do, and how, during intercourse.

However, as far as I am aware, none of the sexual aids heretofore known provide a means by which one partner can control the hip movements of the other during intercourse. Nevertheless, in particular positions, and at specific times, one partner may desire that the other increase their hip movement in order to provide better, deeper or faster penetration.

The current invention therefore provides a device for wearing during sexual intercourse to facilitate this. It consists of a band for fastening around a person's waist and bands for fastening around their thighs. The waist and thigh bands are connected, providing a handle whereby good leverage can be obtained and the wearer's partner can control the wearer's hip movements as desired.

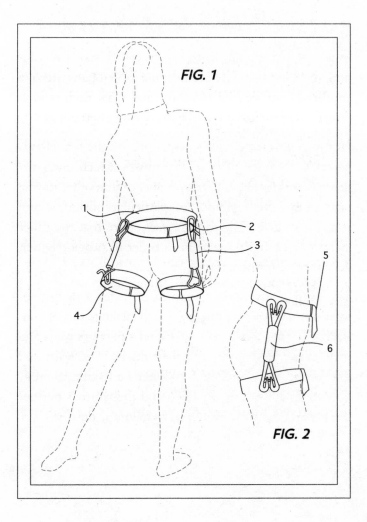

FIG. 1

FIG. 2

Audio-enhanced Vibrator

Play back personalised messages or take notes while using your vibrator.

The present invention concerns a hand-held vibrating device, commonly termed a 'vibrator', used to enhance the pleasure or satisfaction of a person during sexual intercourse or orgasmic therapy.

As well as its general vibrating function, the invention includes an audio-recording and playback system which can be used to record personalised messages or sounds before, during or after sex. The messages or sounds can be used to reduce stress. For example, pre-recorded complimentary comments can be played to generate a romantic mood, or comedic comments can aid in releasing inhibitions.

Further, the vibrator of the present invention can be a convenient device for making notes on non-sexual subjects or ideas that occur during vibrational interplay. Should an interesting thought or idea arise, the user can easily record it without having to interrupt themselves by looking for a pen and paper.

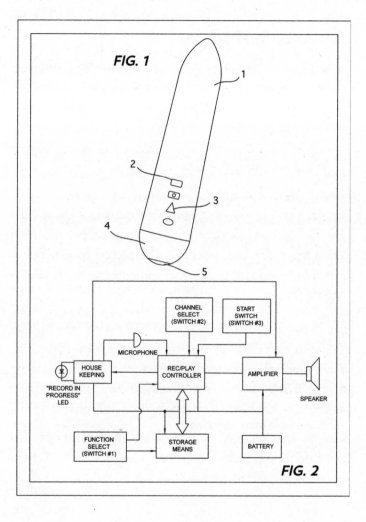

FIG. 1

FIG. 2

Condom Size Selector

Always get the right size condom by taking an accurately scaled photograph of your penis.

Although men's penises come in a variety of shapes and sizes, most commonly available brands of condom do not. Furthermore, marketing shows most men do not know the actual size of their erect penis, and therefore would not know which size condom to buy when presented with a choice.

Self-measurement using a string and ruler is difficult and can lead to inaccurate results and poor-fitting condoms. A need therefore exists for a method to accurately measure the penis. This need is met by the present invention.

The system runs as an application on a handheld electronic device containing a built-in digital camera. First, the user selects a common item of known dimensions. This can be a credit card, a compact disc or a dollar bill. The item is then positioned near the erect penis, and an image of the penis is then taken which is dimensionally scaled with reference to the measurement standard.

The application then calculates the dimensions of the erect penis and links the user to a retail site that sells a condom of the correct size.

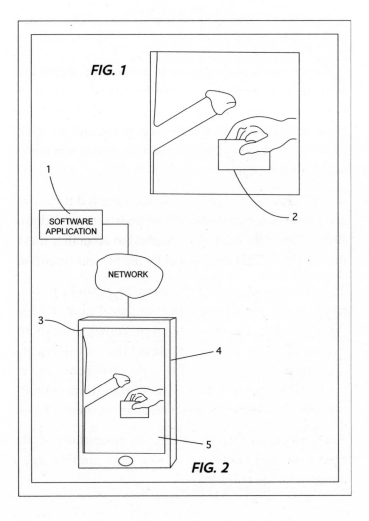

FIG. 1

1

SOFTWARE
APPLICATION

NETWORK

3

4

5

FIG. 2

2

Discreet Stimulation Apparatus

Disguise your vibrator as a teddy bear.

This invention relates to a discreet sexual stimulation apparatus which is designed to be completely inconspicuous when displayed in a typical bedroom or family room. It is primarily for use by women to heighten their personal satisfaction during masturbation or sexual intercourse.

The apparatus is presented in the form of a children's stuffed toy, such as a teddy bear, which contains a vibrator to sexually stimulate the user's genitals or other erogenous areas. The vibrating mechanism itself is positioned within a semi-rigid portion of the toy, such as the mouth, tongue or nose, so that the vibrations are most intensely felt when placed at or near the genital area or erogenous zone. The vibrator is activated by using controls which can be located in the arms, legs or ears of the soft toy.

The present invention thus allows the user easy access to a sexual aid device in the privacy of their bedroom. It is also suitable for use in a television or family room or while travelling as a passenger in a car, where it may be enjoyed unnoticed by others.

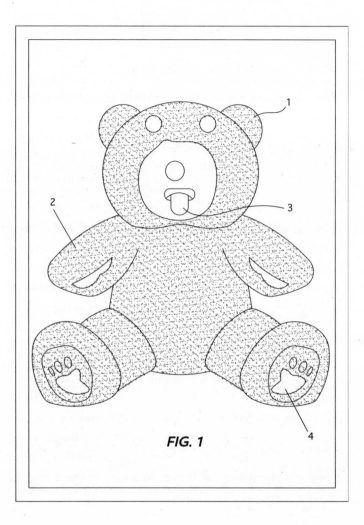

FIG. 1

Condom-warming Radio

Enjoy your favourite radio show while keeping your condom warm and toasty for later.

The current invention relates to a condom apparatus, and more specifically to an innovative condom warming and radio apparatus.

The invention is aimed at making improvements to prior condom devices by providing a condom-heating chamber within a radio structure. The chamber is configured to enable the warming of condom packages in geographical areas with cold, wintry temperatures. This overcomes deficiencies not heretofore addressed in existing devices relative to such areas.

To achieve this end, the invention provides a clock radio housing with two overlying chambers so that the lower chamber directs heat to the chamber above. The upper chamber is configured to hold a row of condoms and includes a push-bar which can be utilised to release condom packages one at a time through a side wall chute as and when required.

It will be appreciated by those skilled in the art that the device is not limited to use in geographical areas with wintry temperatures but may also be used in other regions of the world.

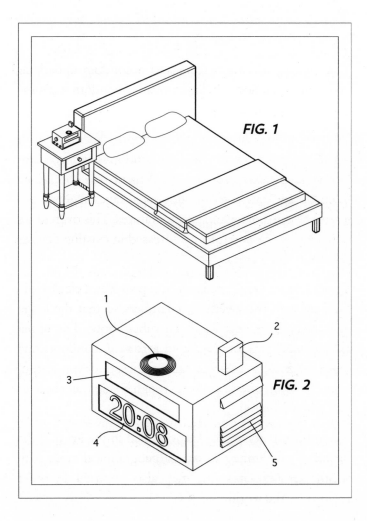

FIG. 1

FIG. 2

Face and Body Condom

Stop your testicles from flouncing and spreading disease.

Even if wearing a condom during sexual intercourse, it is not uncommon for a man to find his testicles coming into direct skin contact with his partner's genitalia. Likewise, if engaged in oral sex, it is possible for his testicles to come into contact with his partner's mouth.

As hard as it may be to discuss, such direct contact with the testicles can lead to the spread of many sexually transmitted diseases, such as herpes, genital warts and HIV, as well as other unwanted maladies, such as pubic lice.

The Face and Body Condom is designed to remedy this problem by not only encompassing the entire penis, but the testicles as well. This is achieved by means of a rubber string at the base of the condom. Said string hooks around to the back of the waistband to secure the testicles in place and stop them from flouncing during sexual intercourse or oral sex.

When engaged in oral sex, the wearer (male or female) would utilise the face and body condom in an inverted manner, using the stretchy waistband as an adjustable headband.

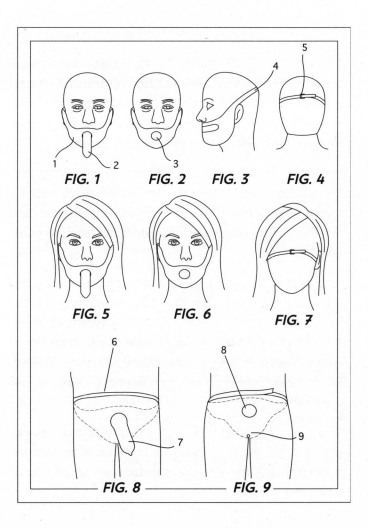

FIG. 1   FIG. 2   FIG. 3   FIG. 4

FIG. 5   FIG. 6   FIG. 7

FIG. 8   FIG. 9

Lap Dance Liner

Make having a lap dance an even classier and more sophisticated experience.

The present invention provides for a plastic pouch and underpant combination designed specifically to provide frictional stimulation to the penis during fully clothed sexual activity such as lap dancing.

The pouch is worn over the sex organs of a man and beneath the underpants, said underpants having a specially adapted elastic waistband to firmly hold it in place. The pouch itself is made of a flexible and elastic material with a slightly rough interior surface. The top edge has an opening which provides access to the hand and wrist of the wearer, so they can accurately insert their sex organs into the pouch.

With the penis positioned in the pouch, the motion of a lap dancer against the outer clothes of the wearer causes the rough, lubricated interior surface of the pouch to rub against the penis in such a way as to simulate intercourse. As long as it is correctly positioned, the pouch captures any fluids released prior to, and during, the lap dance act, thereby facilitating clean-up afterwards.

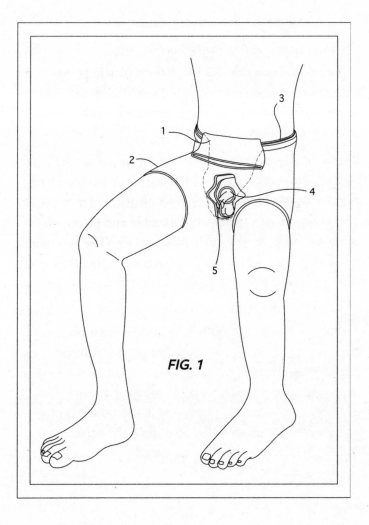

FIG. 1

Safe Sex Brooch

Let everyone know you are carrying a condom.

The recent upsurge in the incidence of sexually trans-mitted diseases (STDs) has increased public awareness of the need for taking health protective measures. At the same time, unmarried people wish to lead the same lives they lived before.

This problem is exacerbated by the fact that men gener-ally fail to carry a condom with them when they socialise, for example at parties. Such social events are important to young, single people, but due to the risk of contract-ing an STD, the idea of sexual contact is often rejected.

The present invention seeks to overcome this problem by providing an ornamental safe sex device that can be pinned to the clothing, highlighting that the wearer is carrying a hidden prophylactic.

The device is in the form of a brooch and includes the indica 'Ss', conveying a subtle message to the viewer, namely 'Safe sex.' A packet containing a condom is posi-tioned in a holder mounted behind the 'Ss' emblem.

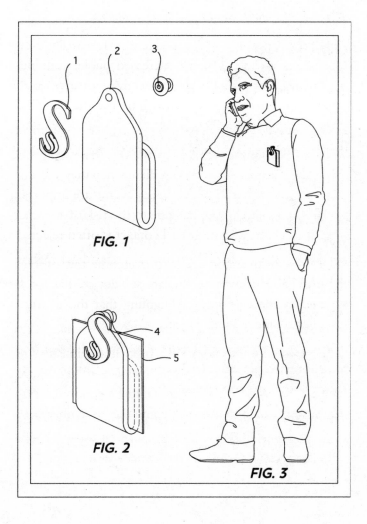

FIG. 1

FIG. 2

FIG. 3

Discreet Sex Chair Device

Keep your sex chair in full view without raising any eyebrows.

Chairs for facilitating sexual congress are known, but have generally been very conspicuous contraptions that are blatantly sexual in nature and function. Since the majority of people do not want such contraptions in their home or office, lest they be embarrassed, these devices have failed to fulfil the needs of the market.

The present invention overcomes these problems by providing a sex chair device that resembles and functions as an ordinary chair. This allows it to be left in open view without causing embarrassment. It can, however, be quickly repositioned for use as a sexual facilitation device when required.

Initially, the device is placed in the upright position, where it can be used as a normal chair for sitting. When it is required for sexual activity, it is simply flipped over to its inverted position. In this position, it can be used to facilitate many forms of sexual intimacy.

The visible wood is preferably oak, or another aesthetically pleasing wood, while the portions covered with upholstery can be made of less expensive material.

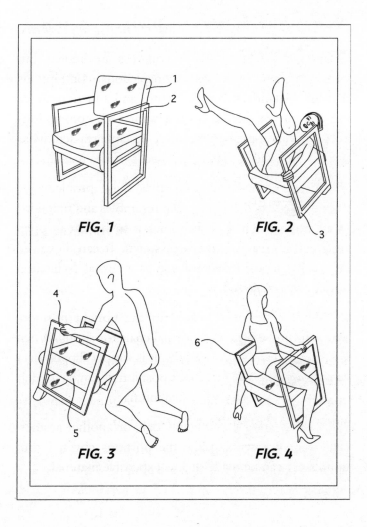

FIG. 1

FIG. 2

FIG. 3

FIG. 4

Man's DIY Erection Truss

Make your own erection aid with tubing, scissors and glue.

Presently, and in the past, several types of products have been used by men to maintain their turgidity during times when male potency is required.

Such devices, however, are limited in how easily they can be customised. There is therefore a need for an erection aid that a man can customise himself while not interfering with the enjoyment experienced either by himself or his sexual partner.

Accordingly, a kit is provided with which a man, either alone or working with someone else, can prepare his own erection truss. The kit contains a length of surgical tubing, plastic fasteners, a position marker, a pair of scissors and a tube of glue.

After fitting, the two plastic fasteners are covered with glue and inserted into each end of the surgical tubing. When the glue sets, the erection truss can then be positioned for the first time, and many times thereafter, around the man's penis and under his testicles.

The truss is easily and quickly cleaned after use. It can then be optionally placed in a writing pen-like container while awaiting another period of successful use.

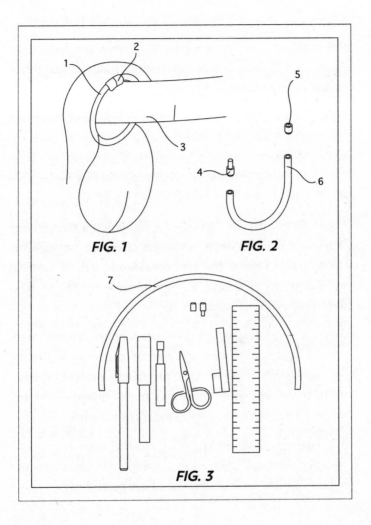

FIG. 1

FIG. 2

FIG. 3

Musical Condom

Simply generate a sufficient level of movement for your favourite tune to play.

Generally, condoms function satisfactorily as a method of birth control, but many people are disinclined to use them because they reduce sensitivity, interrupt copulation and are tiresome to wear. It is therefore the aim of the present invention to provide a device which will encourage condom use by providing entertainment and amusement to the user and their partner.

In operation, the device is placed on the user's penis. During sexual intercourse or other activity, the contact between the genital areas of a couple will create sufficient force to engage the electrical circuit. Power will then pass from the battery to the sensory unit, which will then produce sounds, for example music or a voice message, and/or flashes of light from an LED.

The music or voice message may be played once or it may be repeated continuously for several minutes. Suitable melodies may include 'The 1812 Overture', 'Ode to Joy', 'Happy Birthday', 'My Ding-a-Ling', 'You Light Up My Life', 'We've Only Just Begun' and 'Yellow Submarine'.

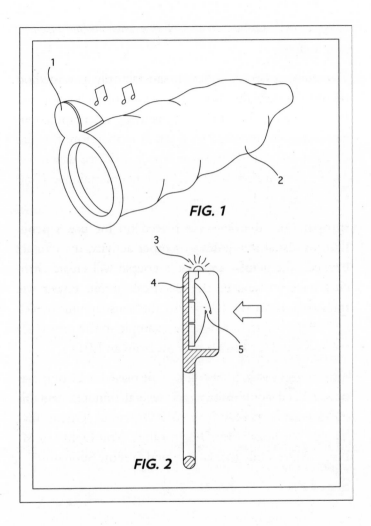

FIG. 1

FIG. 2

Safer-sex Garment System

Ensure really safe sex by providing safer-sex information in your underwear.

The number of people infected with sexually transmitted diseases (STDs) has increased dramatically over the last decades, putting the need for safer-sex aids at a critical level.

The current invention is a direct response to this need. It is provided in the form of a kit that includes a specially designed undergarment made from absorbent material with an opening in the crotch area to allow for penetrative sex. It also includes a number of pockets to hold various sexual aids, such as condoms, lubricants, gels, antibacterial cleansers, scented oils, lotions and safer-sex information.

This information may be presented in the form of a leaflet or brochure detailing safer-sex practices, sexually transmitted disease statistics, symptoms of STD infection, STD testing information and information on the correct use of condoms.

Although the pockets are placed on the interior of the undergarment, pockets may also be attached to the exterior of the undergarment.

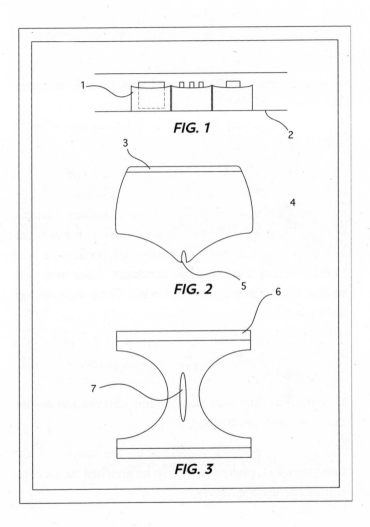

FIG. 1

FIG. 2

FIG. 3

Sexual Device with a Smoking Pipe

Smoke a pipe and suck your partner's penis at the same time.

The present invention provides a novel aid which is designed to prolong the erection of a penis while simultaneously facilitating smoking-related activities.

The device features a penile band, often known as a cock ring, and a mouthpiece with a chamber for holding a smoking pipe. An inhalation opening situated near the tip of the penis directly connects to the pipe-holding chamber.

In use, the penile band is first stretched and then placed around the shaft of the erect penis. This traps blood inside the penis, helping to preserve the erection. At the same time, a smoking pipe can be inserted into the holding chamber so that a partner can smoke while simultaneously performing fellatio or other sexual activities.

The sustained pressure around the penis created by the band allows the wearer to maintain his erection while the partner enjoys the pipe. As will be evident to those skilled in the art, multiple partners can utilise the pipe in turn.

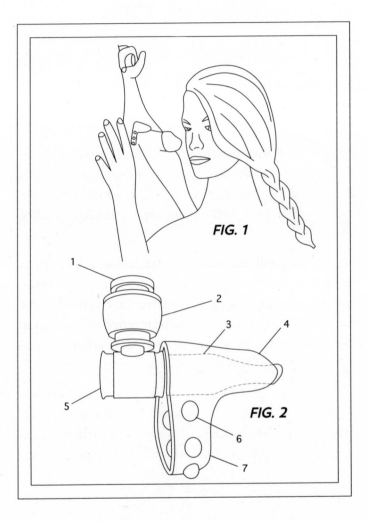

FIG. 1

1

2

3

4

5

FIG. 2

6

7

Acknowledgements

First, I would like to thank my agent, Piers Blofeld, for his advice and support. Piers immediately saw the potential in the mass of material I originally showed him, but just as importantly told me exactly how I needed to present it if I wanted to find a publisher. Thankfully I had the good sense to listen.

Anna Mrowiec commissioned the book for HarperCollins. As senior editor she also oversaw the entire content, production and design. As a result, the book is far better than I had imagined when I first began compiling the material. Alongside Anna, Simon Gerratt managed the project and ensured it met very tight deadlines, Dean Russell co-ordinated the design stages, Rajdeep Singh skilfully redrew all the illustrations, Ellie Game designed the fantastic cover, Alan Cracknell brilliantly produced the book itself, copy-editor Lizzie Henry substantially improved my own editing and proofreader James Ryan picked up on the many mistakes I had missed. Additionally, my long-time friend Ben Wetz cast his expert eye over various early drafts of material, helping me distinguish the brilliantly bad from the simply bad.

For everything else, I would like to thank my family, Lucy and Katherine.